	Cloth price	$8.95
	paperback	$6.95

A SINCERE PRESS PUBLICATION

HOW TO USE SOLAR ENERGY

by William L. Ewers

PUBLISHER
SINCERE PRESS, INC.
BOX 17599
TUCSON, ARIZONA 85731

ISBN   0-912534- 17-6 Library cloth binding
ISBN   0-912534- 18-4 Paperback trade
© Copyright 1976 by William L. Ewers
ALL RIGHTS RESERVED   This book may not be reproduced whole or in part, in any form, without written permission of the author.

PRINTED IN THE UNITED STATES OF AMERICA

Library of Congress Cataloging in Publication Data

Ewers, William.
   How to use solar energy.

   Includes index.
   1. Solar energy. 2. Solar heating. I. Title.
TJ810.E8   621.47   76-17058
ISBN 0-912534-17-6 lib. bdg.
ISBN 0-912534-18-4 pbk.

CONTENTS

Foreword		Page 4
Chapter One	What is Solar Energy?	Page 5
Chapter Two	How can you use Solar Energy	Page 21
Chapter Three	Heating your swimming pool with Solar Energy	Page 37
Chapter Four	Solar Water Heaters	Page 47
Chapter Five	Solar Cooking-Solar Still	Page 70
Chapter Six	Space Heating & Cooling	Page 81
Index		Page 95

Original Artwork by SHERRY LAWRY YOUNG

1945847

Other books from Sincere Press:
    Sincere's Sewing Machine Service, William Ewers
    Zig Zag Sewing Machine Service, Wm Ewers & H. H. Kenaga
    Sincere's Mini-Bike Service, Wm Ewers w/ Irv Charles
    Sincere's Bicycle Service Manual, William L. Ewers
    Home Cleaning Guide, Barbara Molle & Irv Charles
    Sincere's Air Conditioning Service, vol #1, Wm Ewers
    Sincere's Air Conditioning Service, vol #2, Wm Ewers
    Staple Gun in Home & Industry, William Ewers
    Vacuum Cleaner & Small Appliance Service, Wm Ewers
    How to Read on or Above Grade Level, Grace Blossom
    Fundimentals of Tailoring, Freddie Ervin
    Home Swimming Pool Maintenance, William Ewers
    Sincere's Lawn Mower Service, William Ewers

FOREWORD

There is one advantage in a layman writing a book of this nature and that is relative objectivity. I say, relative, because it wasn't very long after I started researching the book that definite opinions began forming. I still can't understand why the United States has taken so long getting started in full-scale solar energy research. In any event, I tried to avoid prolonged interviews with forces that could sway me one way or the other.

After the research and writing is over, there is one point that is primary, and that is that there is sufficient information now available for a full scale utilization of solar energy for heating water, (both domestic and swimming pool), and space heating and cooling. Although there is a cost factor involved, most units now available are within the price range most Americans can afford. There are variables which can influence a given situation (i.e. location, how elaborate the system may be, utility rates, amount of financing that is available, zoning laws, aesthetics) but the fact remains that the technology is here, and now.

And finally, it appears that the government is not going to become involved at levels earlier believed possible. If solar energy is to become a viable alternative to fossil fuel energy, the government must lead the way. When that happens, private industry will join the program and we will have travelled a long way toward energy independence. Let us all pray that day isn't too far away.

William L. Ewers

CHAPTER ONE

WHAT IS SOLAR ENERGY?

The first item of business is just what is meant by the term "Solar Energy", and secondly, how the term will be used in this book. In the technical use of the term, solar energy is the electromagnetic radiation produced on the sun and radiated over 93,000,000 miles, at a speed of 186,000 miles per second, to all of us here on earth. The radiation is the result of on-going nuclear reaction on the sun's surface and it comes to us on earth as sunlight as well as ultraviolet and infra-red rays. The sun provides, both directly and indirectly virtually all energy used on this planet, and in that context, must be considered the basis for all life. Chemical reactions stimulated by the sun's rays formed the fossil fuels(coal, gas and petroleum), millions of years ago. Regardless of religious or scientific application, the sun has been and remains, the prime factor in everything that has or will transpire on earth.

Most everyone has heard of the ozone layer in the atmosphere and how gasses from aersol cans, along with supersonic flight and other man-made substances are destroying it, but what many people do not know is why this layer is so important to life on earth. This is the layer that filters, scatters, and reflects, harmful portions of solar radiation before it reaches us. Additionally, the atmosphere absorbs most of the ultraviolet radiation that would be most harmful to life on earth. Sunlight that finally reaches us is the source of energy for plants, climatic changes, and ocean currents. This is the energy we know as solar energy and is the power source for the projects covered in this book.

Another indirect source of the sun's energy concerns it's relation with the atmosphere which produces winds. Wind has been utilized as an energy source for centuries and is currently enjoying a rebirth. Invention of more efficient storage batteries has reopened the use of wind chargers, especially in rural areas, and among the alternative lifestyle people. We had an old Wincharger on our farm in North Dakota, during the 1930's before REA(Rural Electric).

One estimate states that enough solar energy falls on the United States every 20 minutes to supply our energy needs for an entire year. The average in near earth space incidence of solar energy is approximately 130 thermal watts per square foot, and that on the ground average in the United States is approximately 17 thermal watts per square foot which results in a average daily energy supply of 410 thermal watt hours per square foot - about twice the amount required to heat and cool the average home. However, when you consider the conversion factor which is around 10% efficiency, there is some difference, but scientists estimate, based on 1969 total energy figures, that all energy consumed could have been supplied by a solar energy incidence on 0.14% of the United States land area.

The problem facing our scientists, and commercial interests involved in solar energy research, is how to collect, utilize efficiently and store that power, with methods of storage the biggest obstacle to overcome. The use of solar energy for heating water, space heating, cooking, food drying, distilling and growing plants in indoor greenhouses is nothing new. In fact, before the widespread use of fossil fuels for heat and power, the sun was the primary source of energy. History tells us that solar furnaces were in use during medieval times and solar energy smelters were in use by the late 1700's.

Solar research of modern times had been plodding along in a rather unexciting manner from the late 19th century, until the recent oil embargo again thrust the energy problem into the limelight. This is true mostly in the United States since solar energy development has been proceeding at a faster pace in such countries as Japan, France, Israel, Australia and Russia. This has been due to the need for low cost power, a very real need for alternative energy sources, and a lack of technology that we take for granted. There have been industrial research projects in other countries since well before 1900, notably a solar-powered steam engine for operating newspaper printing presses in Paris, and a solar-powered irrigation system in the middle of the desert near Meadi, Egypt, around 1913.

Solar water heaters were used in Florida starting in the 1920's, and continued in popularity until the early 1950's before starting a decline which was brought about primarily due to the availability and low cost of natural gas. That is probably the reason more research wasn't undertaken, since about the time some of the aforementioned projects were started, the availability and cost of fossil fuels was just too much competition.

The sad fact is that this country of almost unlimited technology is lagging behind countries such as Australia, Japan, Israel, France and the USSR. Solar water heaters are manufactured commercially in all of these countries, but only on a limited scale in the United States. In fact, the solar water heater is standard equipment in many places, especially Australia.

Japan is not only a leader in development of solar water heaters but in the development of solar cells as well, where they are well into a multi-billion dollar program known as "The Sunshine Project", which is projected over a twenty five year period. Some of the Japanese companies involved in solar energy research and development as well as commercial manufacturing include Nippon, Sharp (electronics people) and the Hitachi company, with Hitachi already marketing in the United States on a limited scale.

Solar systems were widely used in Israel after they gained independance, but as the country developed, utility companies produced low cost energy and for a time solar energy was relegated to a minor role. The continuing Middle East turmoil has changed that.

The Russians use solar cells in their spacecraft program and have been involved in a research and development program in Tashkent for many years. This project is for generating electricity on a big scale.

And so it goes around the World. Smaller, developing nations, as well as many well versed in technology, now realize that use of relatively free solar energy is a must. The threat of another oil embargo is always a possibility and we found, much to our chagrin, that even might nations are not immune.

One big area of R & D (research and development) in this country has been in the use of solar cells in spacecraft and satellites. Most of the spacecraft circling the globe are powered and their batteries are recharged by solar cells. The biggest obstacle to the use of solar cells for domestic application is the cost, which at this point is prohibitive. However, several companies and research firms are experimenting with "growing" silicon crystal (which is the base substance for most solar cells) and if mass production is around the corner, costs will come down. When this will happen is anybodys guess, but if more federal funding is forthcoming, the time span will be greatly reduced.

Most of the programs in other countries is funded, or at least underwritten, by their governments, but in this country, until very recently, most research has been with private money and therein lies one problem. Private industry-big business-has been, for the most part, unwilling to enter a somewhat chancey field in which the government has been unwilling to advance substantial funds. There are exceptions to this with companies such as PPG, Revere Copper, Honeywell, Westinghouse, Arkla, Olin Brass and Motorola already marketing components of one type or another specifically for this industry.

The oil embargo and subsequent fuel shortage forced a change in government thinking and increased activity in the area of alternative energy. Among the considerations are nuclear energy on a grander scale, solar and wind power, methane gas and a journey back to the old standby, coal.

In the period of two years, R & D funding jumped from the million dollar a year range to the fifty million dollar range. Belatedly, the Congress passed two meaningful laws in 1974. The first was a two stage, five year program for research and development of combined space heating/cooling systems, and the other was an R & D energy program related to the alternative sources mentioned above. The problem was, even with the increased outcry against nuclear power that solar energy was relegated a back seat. The chairman of the Atomic Energy Commission at that time, Dixy Lee Ray, continued to push nuclear research and cut back solar energy research, and solar energy development was also a section of that commission. (Editor's Note: As recently as February 1976, in a speech to the Southern Arizona Chapter of the Electrical League of Arizona, Ms. Ray couldn't conceal her position on solar energy when she stated there would be no significant impact from that energy source until after the year 2000, and that economical feasibility was at least 25 years away. The thrust of her speech was increased emphasis on nuclear power to avoid dependence on foreign oil). We can hardly argue with the need for independence, except to state that increased research into solar energy programs could eventually reduce those gigantic costs she refers to when talking about generating electricity from solar energy. She is, of course, referring to generating electricity on a large scale, but her adversary position-regarding solar energy research-is scarcely concealed. She made the point, and validly - that utilizing solar energy was like trying to harness 100 million fleas then teaching them to all jump in the same direction at the same time, while harnessing nuclear energy would be the same as harnessing one elephant, relatively speaking.

We still believe that additional R & D money is needed to solve the problems and gain a parity with existing energy sources. If an equal amount of money was spent in solar research as has been spent in nuclear energy research (just the peacetime expenditures-not bombs) we would have a viable program within five years, ten at the most.

Ms. Ray is no longer chairman of the AEC so she was not speaking from an official position, but represents a good example of thinking on that level.

Fortunately, there are several senators and congressmen who are working within the system to bring about change and appropriate more money for research. Senator Mike Gravel of Alaska has been a prime mover in this area, as have numerous other men. There was a rare showing of solidarity in passing the Solar Heating and Cooling Demonstration Act of 1974. In brief, this act earmarked 50 million dollars to buy 2,000 solar-powered heating units, and another 2,000 heating and cooling units -in family dwellings - to be tested in various parts of the country.

The two federal agencies most influential in handling the 1974 act are NASA and HUD. One of the programs is the several million dollar grant administered by HUD (Housing and Urban Development) which will dispense the funds for new housing units utilizing solar units, as well as full payment grants to builders or individuals who add solar energy packages to homes. The grants are for the solar energy package only.

The federal agency most directly responsible for coordinating all alternative resource programs is ERDA (Energy Research and Development Administration). ERDA was originally directed to handle the Solar Energy Information Service with a two million dollar grant to fund a NSF (National Science Foundation) study and to establish a new Solar Energy Research Institute somewhere in the United States. Many states are trying for this Institute but the site has not been chosen. ERDA is also working with HUD in the coordination of the residential programs mentioned previoulsy. NOTE: ERDA has a technical information service available to anyone requesting one. Mail your request to ERDA, Technical Information Center, Post Office Box 62, Oak Ridge, Tennessee, 37830.

There has been a definite increase in the utilization of solar energy in the area of public buildings, especially in schools, public libraries, fire stations, and other office buildings. The Forest Service has erected public outhouses in Montana with solar energy as the power source.

In my hometown of Tucson, Arizona, there are several projects currently underway including a fire station and recycling plant. The projects are jointly funded by the City of Tucson and University of Arizona.

FAUQUIER HIGH SCHOOL- WARRENTON, VIRGINIA

Page 11

Several projects involving the use of solar energy in public schools have been completed. These include a project in Minneapolis, Minnesota, and others in Atlanta, Georgia, the Timonium school in the state of Maryland (see illustration below), and the Fauquier School in Warrenton, Virginia (see illustration previous page). This would appear to be a natural application, since the solar energy system could also serve as a learning center.

The biggest project to date has been the 1.5 million dollar installation at New Mexico State in Las Cruces, New Mexico. The unit involved is an agriculture building and the system covers all solar systems (i.e. heating, cooling, heating water, etc.).

TIMONIUM SCHOOL - MARYLAND

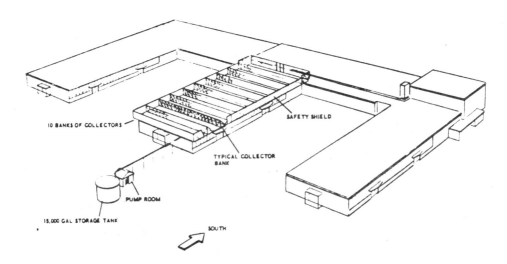

It does appear that a viable solar energy program is finally underway, roadblocks and all. If media coverage of solar energy was an indication of progress we would be in a solar energy boom. Hardly a day passes without some mention of the subject both in newspapers and broadcasting/television. All of this, combined with announcements by governmental agencies, scientists, industrial firms, and self-styled experts, no doubt tends to confuse, rather than enlighten the average person. We'll try to reduce solar energy concepts to a basic level of understanding and clear up some misconceptions at the same time. Additionally, while reviewing several applications of solar energy, we will also delve into some of the legal and moral aspects of using, what appears on the surface at least, a virtually no-cost source of power.

There is one very real question to be answered at this point, " Is Solar Energy a viable alternative power source?" Or is it merely a fad, a glamour subject meant to conjur up dreams of energy independence, and on the darker side, a get-rich-quick thing destined to enrich a select few and sucker in millions of unwary investors or users. Indications are that solar energy is perhaps both of these, and more, since in anything new or faddish (not that solar energy, per se, is new) there are many varied and controversial sides to this story.

This probably stems back to the Florida Solar Water Heater industry, which was a mixture of success and failure, to the early experimental house at M.I.T. and on through a parade of good, and not so good. One of the pioneers and Holder of many solar home patents is Harry Thomason. His original solar house in the Washington, D.C. area is still in operation, as are several other houses he has built. His concepts have proven most valid and through the years since that first house of the 1950's, he has continued to modify and improve on his original design.

Any discussion of solar energy always includes comparative costs. First, the comparatively high cost of initial installation (whether on an individual basis or for generating massive amounts of power), and secondly, the free sunlight for future operation. It boils down to X number of cents per kilowatt hour compared to public utilities. Critics of the high initial cost conveniently forget the massive R & D costs associated with existing power plants including power lines strung all over the country. Money spent in nuclear research alone would absolutely boggle the mind. The total outlay for hydroelectric and coal fired plants is certainly not a small figure. All things are relative and if we continue using the finite resources we now take for granted, future costs for a crash program in solar energy may be dear indeed. If the same dollar value spent in fossil fuel and in nuclear research since World War II had been spent in solar energy research we would now be enjoying lower utility bills. And be well on our way to energy independence with a greater reserve of fossil fuels for future generations.

One of my favorite people, President Harry S. Truman, stated to The Materials Policy Commission in 1952, "Efforts made to date to harness solar energy economically are infintesimal. It is time for aggressive research in the whole field of solar energy, an effort in which the United States could make an immense contribution to the welfare of the free world." Prophetic? Twenty Years later ? I certainly think so, and fortunately, it's still not too late.

As mentioned previously, cost is definitely an important factor, since the initial outlay in any type of project is somewhat more than for a comparable system utilizing electric or fuel powered sources. However, we now arrive at the other side of the coin, that of operating costs, since after the initial outlay the energy source is free. A comparison of maintenance costs should find that a solar system costs much less to maintain.

Let's reduce this hypothesis to an individual level and use as an example a average swimming pool. The cost of installing a simple pool heating system would be comparable except for the addition of solar collectors(depending upon power source, size of pool, and the location), but operating expenses would soon bear out the wisdom of the solar system. Referring back to location, it would be more valid in the southern section of the country but is valid everywhere.

The pool blanket solar heater can keep a pool warm for most days of the typical pool season for just a fraction of the cost incurred by a fuel-fired system.

A similiar parallel can be drawn for space heating/cooling and for heating water. The initial cost is higher, operating cost is lower. The period required to amortize the cost of installation is generally the deciding factor.

We've been looking at solar energy pretty much from an individual viewpoint, but what is the potential for widespread use. Is solar energy really a viable alternative? Will huge solar farms or other concepts for mass producing electricity ever come to pass? That is a good question. Doctor's Aden and Marjorie Meinel, from the University of Arizona have long espoused building huge solar farms set in remote desert areas. The farms would consist of acres of collectors which would transfer the energy to generating plants and finally electricity, which could be transported over existing power lines. There are many variations to this concept but the capability for performance isn't that far away. It would require cooperation of the utility companies and recently they have become more aware of this energy potential and seem to be doing something about it.

The availability of low cost energy has changed this country and utility companies have had no small part in this. The simple life of our youth - some people say harsh life - has now become totally unacceptable, except for the small migration of some back to what is called alternative lifestyles, and we've become prisoners of automatic appliances and labor saving devices. Most of which are run by electricity or fuel.

You might say the American public has been subjected to the most intensive brainwashing known to man. In may ways I've come to an understanding with the dissenters and dropouts of the 1960's and early 1970's. They were right, we have become very materialistic and the selling job was - and is for that matter - second to none. It doesn't make sense for utility companies-both publicly and privately owned - virtual monopolies - to constantly exhort us to buy more air conditioners and appliances. The money spent in advertising alone by utilities, could bring us a viable solar energy program.

When the energy crunch arrived in the early 1970's, they suddenly switched to the role of public savior by urging us to conserve energy or we would run out of fossil fuel reserves. Did they follow their own advice? Rarely, and the result has been skyrocketing utility rates. The situation is so critical in some areas utility bills are higher than mortgage payments. And finally, the ultimate irony of all time. When reduced usage resulted in reduced revenues, these public spirited bastions of our society raced to regulatory agencies for permission to raise rates because profits were down. And in the case of investor owned companies, the stated reason was, they needed more profit to keep a good investment image, which means increased profits all around.

It would be a tremendous plus to solar energy research if the utility companies used some of those profits to aid the cause. There are indications this may come to pass.

In any event, the use of solar energy on any level will have the effect of diluting the strength of some of the giant utilities and saving our fossil fuel reserves at the same time.

I don't think it's necessary to return to the days of my youth when only the cities were blessed with abundant electricity, and natural gas usage was still growing, but I think it's possible to live a good life on far less than that to which we've become accustomed.

There must be a happy medium and I'm convinced that position can be attained through the widespread use of solar energy. We have alluded to the need for large sums of money to get any kind of full-scale solar energy development off the ground, and that fact cannot be emphasized enough. As has been the case with gas and oil exploration, some kind of a government subsidy is a must, at least to get the ball rolling. We also know that once government money is involved, private industry will not be far behind. It's a fact that has been proven many times over, not just in the gas and oil industry, but in health care, aerospace, and others too numerous to mention. In my opinion, there comes a time when some of our tax dollars should return to assist us on more than a minimal basis.

There should be no hint of partisan politics, in fact, the opposite has proven true with legislators from both sides of the aisle joining in a concerted effort to pass solar energy legislation. All that now remains is a joint effort by the industrial community and utility companies in concert with government to get the ball rolling. Utopian? Perhaps. At least in the foreseeable future, but hopefully, public pressure and government sponsorship on pilot programs can change the situation.

A program under sponsorship of ERDA calls for the completion, by 1980, of a pilot 10,000 kilowatt plant somewhere in the United States. Total funding of Arizona's proposal will be seventy million dollars; fifteen million from the private sector and fifty five from the government. The private money would be for construction of the auxiliary, or backup section of the plant, while the government will finance the solar generating equipment (including a collector system of gigantic proportions).

In Arizona, Tucson Gas & Electric along with Arizona Public Service and the Salt River Project will bid on the project in competition with utility companies from Florida, Southern California, Texas, New Mexico, and probably others.

It's a great idea and should give researchers an opportunity to see firsthand if large scale systems will be econimically feasible. The 10,000 Watt plant is small by most standards, and present plans call for just one such project. And to give our local utility, Tucson Gas and Electric a pat on the back, they are allocating their entire research budget for the next five years, to the project.

It now appears solar energy development is underway, perhaps we can go a step farther and expend some of our resources into an area of genuine need, low income families of the United States.

One of the tragedies of increased energy costs is the effect it has on the low income family and elderly. They are having increased difficulty paying climbing utility bills. Since it's almost impossible to bring about any equity through the companies, and cheap energy seems to be gone forever, maybe we could work something out in another area.

Why then haven't we looked to the sun before? As we now know, solar energy is nothing new. Why haven't we used this source as a vehicle for improving the lives of our poor, not only in this country but the World as well? We could use existing technology to build basic solar systems, on a retrofit basis, on many of those homes. The units wouldn't have to be too sophisticated. As a start, we can install water heating and space heating systems, with perhaps cooling at a later time. Think of the countless millions who could go to bed each night with a hot bath and heat throughout the home.

A basic system, with a small array of collectors, connecting pipes to a storage tank surrounded by rocks could improve the lot of many people. And when used in concert with existing systems, a basic system could reduce utility bills substantially. It appears to me, that if the leaders in Washington want to assist the less fortunate, this would be a great place to start.

The bureacracy for handling such a venture is already in force. In fact, a program similiar to Model Cities in which people themselves defray a portion of the cost could be instituted, and existing offices within HEW(Health, Education and Welfare) could coordinate the effort. This would eliminate the creation of yet another bureau.

Under my plan, a basic solar system would be utilized, with the materials subsidized by HEW, but the homeowner would furnish the labor. By using less costly materials, such as corrugated sheet steel and black PVC pipe, along with rocks(which are everywhere in abundance), and metal tanks which could be obtained from many places, including U. S. Government surplus depots, a relatively inexpensive system could be constructed.

It appears there is merit to the plan, and we have the ingredients for a successful program. There is an abundant energy source - the sun - and a huge potential market, the low income people of the World. We could accomplish two goals; first, reduce reliance on conventional energy sources, and secondly, upgrade the standard of living for millions of people. And since the concept is a proven one, we would not be using the people as guinea pigs. It could be the ultimate situation, namely reducing the plan to the simplest of terms, by building the program from the ground up. It will prove that solar energy is a viable alternative to fossil fuels, and that by beginning with a basic unit, any home or business can benefit. Since erection of the initial unit is where the money must go it's simply a matter of aiming for that pointing and taking off.

One final word about low income dwellings. An additional study could be made to determine how the homes are insulated and since many will be sadly lacking in that area, perhaps a plan for insulating all homes of need could be integrated into the total package.

A significant side effect would be a dramatic reduction in illness and the subsequent cost of doctors and hospitals, which would result in less of a drain on public health services.

Food preservation is another area for consideration. Many poor families lack proper nutrition they can't preserve those foods that tend to improve their diet. A solar food drier is a very basic unit that isn't too costly and yet could be a tremendous asset to the low income family (the food drier is gaining in popularity especially in alternative lifestyle families).

Another movement that is gaining ground across the country is the use of tax credits for builders of solar units. Included in the many tax bills under consideration - or already passed - are income tax credits, sales tax or property tax credits. Indiana was the first to pass such a law, and at this writing nine others have followed. They are Colorado, Montana, Maryland, North and South Dakota, Florida, Texas, & New Hampshire; Arizona is in the process of passing such a law (summer 1975). The Indiana plan is for property tax deduction up to $2,000 annually.

There are other variables to be worked out; one very real problem is with existing zoning laws. Zoning laws should not be a factor in commercial or industrial projects, since there is generally a wider allowable variance in those areas, but residential zoning can be a different story. I think most everyone will agree that aesthetics is not one of a solar energy unit's long suits. We've already had one case in Tucson, involving a upper middle class neighborhood in the Catalina Foothills. In this case, which went to court, the solar unit construction was denied.

My advice is, check local zoning laws thoroughly, and this applies to retrofit units as well as original equipment installation.

The question posed earlier regarding "free sunlight" is a definite consideration. It will apparently face an early court test in California where state legislators are currently wrestling with the problem. This is more than a superficial problem because it involves financing solar energy units. Financial institutions aren't about to loan money on any venture that isn't tied down legally.

The questions involved are: 1. Who owns sunlight? 2. How much are the sun's rays worth? If people had the rights to sunlight which strikes their property and could sell or transfer those rights, there would be greater value in any solar energy program.

A second and definitely moral as well as legal question is whether a neighbor might be allowed to plant trees or erect a structure which would blot out a portion of the sunlight, rendering a solar system inoperable, or at the very least, diminishing it's capability. This would be more applicable in the building of a skyscraper which could alter the sunlight from several buildings. This right of sunlight is a question that must be answered before any viable program can get off the ground.

Earlier in the chapter, I referred to the darker side of the solar energy program. It's sad but true that any new industry seems to attract shady operators with fast buck schemes. If this program is to be successful, the industry leaders must establish guidelines and standards to insure product effectiveness and safety. This must be accomplished soon to insure any success of the industry. There must be a compatibility between system components and the buildings to which they are attached.

If the industry establishes a good reputation from the beginning, the chances for success are great. But if any shoddy practices are allowed there could be rough times ahead. The American consumer is much more aware of bad business practices as in years past, so if the word gets around that the solar energy industry is a rip-off, it may never get off the ground.

Mass production of solar components will bring more larger manufacturers into the business and this should have a stabilizing effect. The reputation of new companies, together with those already in the business, should enhance public acceptance of the industry.

The industry must also make it quite clear that solar energy is not a panacea for all energy problems. They must show the warts as well as the beauty of the program. There are certain limitations such as insufficient insulation on existing buildings and other considerations such as buildings of sub-standard that will not adapt to a retrofit solar energy system; this must be pointed out to avoid a later problem. The worst possible approach would be overselling.

CHART BY HONEYWELL

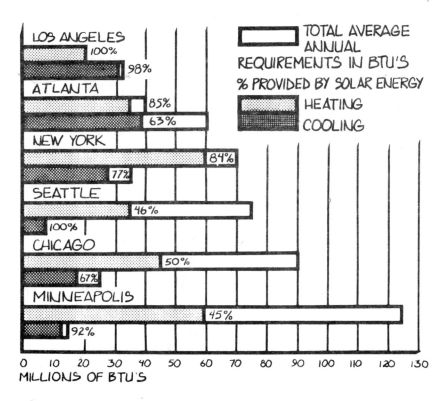

Regional Climatic Classification for the Heating Season
(From November through April)

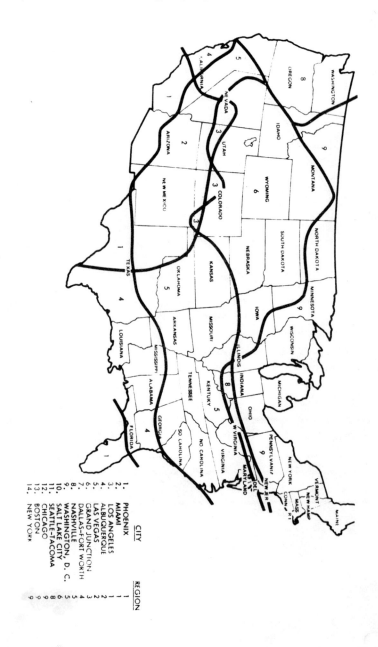

CITY	REGION
1. PHOENIX	1
2. MIAMI	1
3. LOS ANGELES	2
4. ALBUQUERQUE	2
5. LAS VEGAS	3
6. GRAND JUNCTION	3
7. DALLAS-FORT WORTH	4
8. NASHVILLE	5
9. WASHINGTON, D. C.	6
10. SALT LAKE CITY	6
11. SEATTLE-TACOMA	8
12. CHICAGO	9
13. BOSTON	9
14. NEW YORK	9

CHAPTER TWO

HOW YOU CAN USE SOLAR ENERGY

It has become quite apparent by this time that any widespread use of solar energy, in the immediate future at least, will be on an individual basis. As the final draft of this book is being completed there are already signs of fishtailing by the government on future outlay for solar energy research on any grand scale.
The search for a multi-million dollar research center has now taken on political overtones and may become the ultimate political plum for the "RIGHT" city or state. Additionally, dialogue out of Washington speaks of locating the research center in an existing facility that once served another purpose (i.e. a closed defense plant or in the case of a recent President Ford speech in Florida, at Cape Kennedy), and with a vastly reduced budget to operate the program.
So it appears that talk which once centered on founding an R & D center, through ERDA, with the express purpose of finding a viable plan for generating electricity on a large scale, is now reduced to smaller pilot projects. And the only advances in the foreseeable future will come from the private sector as individual projects. It will probably be a building here, a cluster of homes there, kind of thing until another energy crises strikes. Then and only then, and no doubt in a mad dash type of program, will the full impact of the government be utilized to get us into full scale development of solar energy generating plants.
That brings us to the book's theme. The do-it-yourself approach to solar energy use. We will delve into the utilization of systems currently available to the average person, and how we as individuals can help alleviate at least part of the energy problem.

We can start with a relatively inexpensive method of heating water or space heating/cooling for new and existing homes or offices. It is possible to return to a time-as was the case in Florida and Southwest from the 1920's to the 1950's - when widespread use of solar water heaters was in vogue, and once again heat water with the sun.

The construction of a solar water heater is well within the capability of the average do-it-yourselfer (as a subsequent chapter will show) or there are contractors in most parts of this country who are qualified to build one for you. As for the rest of the world, the same is true because solar water heaters are widely used, especially in Australia, Israel and Japan.

The Florida solar water heater industry faded away for two basic reasons:
1. The systems, especially water tanks, were not constructed to last for long periods of time. Improvements and modifications since that time have eliminated most of those problems as we shall see in later text.
2. The sudden availability and low cost for natural gas in the area. We are all too familiar with the cost of natural gas today, and of course, the relative unavailability of same. The situation is so critical in some areas that utility companies no longer accept new gas hookups. Every winter finds more and more curtailments to industrial plants, and as the shortages continue, the homeowner will be next. Several states are preparing legislation barring new gas hookups for heating swimming pools. (Authors note: This is already taking place in California and Arizona and other states won't be far behind). However, as we shall see in chapter three, the marriage between solar energy and heating a swimming pool is a natural.

The advanced technology, coupled with a desire to re-examine the potential of heating water with the sun has returned the program to some prominence. The system can be installed as the primary source - the addition of controls makes this feasible - or as a backup to an existing fuel fired system.

The solar water heater hasn't really changed too much from those early units in Florida. There have been modifications and some improvements, but the basic system of flat plate collectors with a water storage tank are pretty much the same. In fact, the flat plate collector has changed very little.

The use of solar energy for space heating and cooling probably has the greatest potential for growth. A typical system utilizes a solar collector in which water, or in other systems, air, is heated, then passed either directly through ductwork to the living area, or to an insulated storage tank or rock storage bin(or combination of both) to an auxiliary supply unit to back up the solar unit as required. The cooling cycle is more complicated and operates somewhat similiar to the gas-fired air conditioning systems. It requires the use of pumps and other controls to convert heat to cooled air before it is circulated to the living area. See the illustration below. Schematic of a typical system using heated water.

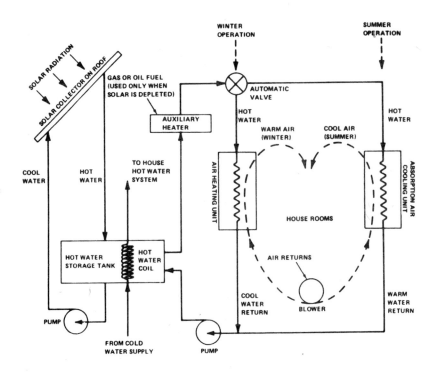

CROSS SECTION OF A TYPICAL FLAT PLATE COLLECTOR

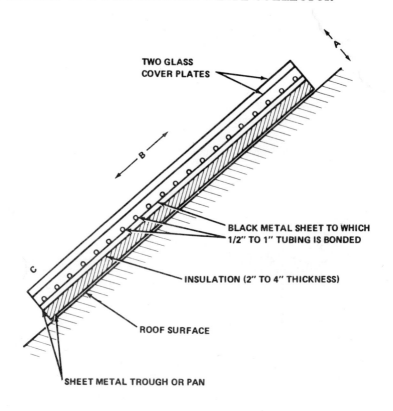

NOTES: ENDS OF TUBES MANIFOLDED TOGETHER
ONE TO THREE GLASS COVERS DEPENDING ON CONDITIONS

DIMENSIONS: THICKNESS (A DIRECTION) 3 INCHES TO 6 INCHES
LENGTH (B DIRECTION) 4 FEET TO 20 FEET
WIDTH (C DIRECTION) 10 FEET TO 50 FEET
SLOPE DEPENDENT ON LOCATION AND ON WINTER-SUMMER LOAD COMPARISON

Page 25

THE SOLAR COLLECTOR

The two most popular solar collectors now available are the flat plate (see illustration previous page and below(a) and the parabolic collector(illustration (b).

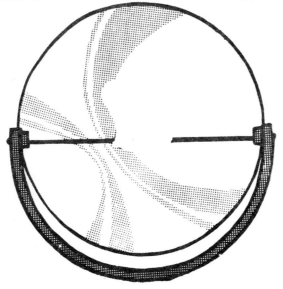

The typical flat plate collector consists of a frame, coils of copper tubing attached to a painted black base, with a pane (panes) of glass on top facing the sun. The copper tubing should also be painted black and soldered to the metal base. An insulation of some kind should be placed between the collector and house roof. There are variations to this including an inexpensive, but not quite as efficient, version utilizing corrugated steel for both the bottom and as troughs for the water, thus eliminating the copper tubing (see illustration), or black plastic pipe which is sometimes used. A clear synthetic material such as polyethelyne is sometimes used instead of glass. It has the advantage of not breaking as easily, but is not as effective as glass.

There are many variations and types of solar collectors using the flat plate concept and we will cover some of these later in the book.

A parabolic collector is generally aluminum or some mirror-like substance which catches the sun's rays to build up heat. This type of collector acts more as a concentrator as the mirrors concentrate those rays toward a receiving component also painted black (such as a tube) and the water is then circulated into the system. The more

COPPER TUBING CONFIGURATION IN A FLAT PLATE UNIT

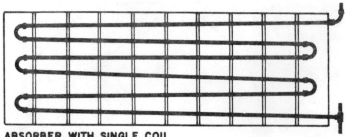

ABSORBER WITH SINGLE COIL

sophisticated versions contain a timing device which tracks the sun across the sky, thereby always catching a maximum of rays. The parabolic collector is seemingly more effective in areas of great sunlight (such as area 1 on the map, page 20), and will probably be the collector type used in the large solar farm arrays.

THE FLAT PLATE COLLECTOR

Construction of a flat plate collector is very simple. The sun's rays penetrate the glass cover(covers) and are absorbed by the black metal base and copper coils (also painted black), which heats the water, or air in some models, as it circulates through the tubes. The same is true for models featuring a corrugated base, or in yet another concept featuring a series of fins affixed to the base. Many designs feature the use of plastic or polyethelyne cover instead of glass. As the sun's rays penetrate the clear cover, the incoming heat is trapped. To put it simply, the glass allows the sun's rays in, but traps the heat which is absorbed by the blackened portion of the collector, which heats the water or air travelling through the tubing to the system. I've avoided using the terms "conduction" and "convection" in this context; they will be covered later in the book.

If this concept sounds vaguely familiar it might well be, because this is the basic principle of greenhouse operation. The difference is that we're using the collector box to heat either water or air, not supply sunlight to plants to increase growth.

Since most applications require more than a single collector, regardless of manageable size, a single collector is one of a series of collectors engineered to handle the requirements for a particular installation. Multiple collectors are joined together by pipes-both ingress and egress, and affixed to desired location on a permanent basis. The following illustration is an example of multiple collectors in place on a school in Maryland. It shows how banks of collectors can be installed. Rarely would a residence require that amount of collector exposure.

Text on how to construct collector modules will be covered in context with the type of unit being covered(i. e. with component needed to construct a swimming pool heater, water heater system, heating-cooling system, etc.). All necessary components for building solar collector modules should be readily available most everywhere in the United States.

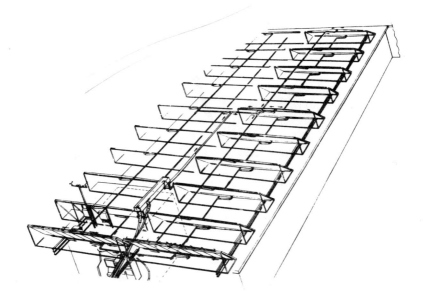

SOLAR CELLS

Solar cells-also known as photovoltiac cells-are one result of the extensive research and development in the space program. There are, or have been, more than 600 spacecraft so equipped. Most spacecraft and satellites are powered by manmade solar cells which convert sunlight to electricity. Solar cells are one of the reasons the space program has been so successful. Without the ability to recharge batteries and power systems in space the program would never have achieved any degree of success. Russia also utilizes this form of generating power in their space program.

The solar cell is a unique invention that converts sunlight into DC (Direct current) electricity to provide energy. It is a solid-state device formed primarily of silicon-there are other substances now used to make solar cells and we'll discuss them later-and yet contains no moving parts. When placed in series, the solar cell can supply energy to anything currently powered by electricity.

The material most used in solar cell manufacture at the present time is silicon, a substance which is found everywhere on earth. And don't worry about stripping the earth of yet another natural resource because silicon is one of the more plentiful elements around. A conservative estimate is that it comprises more than one fourth of the earth's crust; it is perhaps our second most plentiful element.

Simple solar cells are used to power small transistor radios, or automatic light controls, in headlight-activated garage doors, and many more similiar applications. More efficient models are used for charging batteries, or buoys, microwave relay stations, or as a power source for machinery and other signal devices.

How do solar cells relate to solar energy research? Solar cells-photovoltiac cells-could well be the answer, a needed breakthrough necessary to make solar energy a viable power source. We know that the most expensive portion of a flat plate unit is the collector array. And of course this is true of solar cells, except more so. It would require reducing the Kw cost to around $500.00 to compete with nuclear energy costs, which means reducing the cost of manufacturing silicon-or other substances such as cadmium sulfide- to a level where array construction would be feasible. There are some signs of significant success in silicon production (Tyco Solar Energy Corporation is perfecting a process for "growing" silicon ribbons as long as 80 feet, at a speed of two inches per minute), but the cost is still prohibitive for the average home owner. This means the use of solar cells for space heating/cooling or heating water, for the average do-it-yourselfer, is still in the future.

If you would like to experiment with solar cells may I suggest you write: Edmund Scientific Company, Barrington, New Jersey, 00807, for their latest catalog. They offer solar cells of varying costs and effectiveness for experimental, as well as functional, purposes.

Another hurdle to be overcome in solar cell usage on any large scale is the percentage of effectiveness, which is currently in the 10 to 14 percent range. Scientists feel that a maximum of at least 25% efficiency for conversion of sunlight to usable electrical charge in a single semiconductor unit, at room temperature, is obtainable.

In reference to the ERDA bulletin "Solar Energy" and using a 10% efficiency factor, a average size home equipped with an array of solar cell panels 20 feet by 30 feet could collect an average of 25kw hours per 24 hour day. This would be sufficient to supply the power requirements for the average family. Larger homes would require larger solar cell arrays.

Use of solar cells to generate enough electricity for mass usage by consumers is somewhere in the future. Some of the concepts currently under study include solar cell arrays on buildings which include integrated convertors for supplying all required energy to the building, central systems erected on the ground with huge arrays of solar cells capable of serving an entire distribution system, and finally, central systems somewhere out in space, which beam power back to earth to central stations and into distribution systems. See the illustration on page 32.

Although most solar cells are made from silicon, there are other substances that have been used in limited quantities. These include cadmium sulfide, cadmium telluride, and gallium orsenide.

Cadmium sulfide is currently being researched and tested on a large scale and the manufacture will probably be the most feasible due to the relative inexpensive cost, once some of the bugs associated with making solar cells from the substance are ironed out.

There are indications that mass production of solar cells is just around the corner. And, as is almost always the case, mass production will mean a cost reduction. This, coupled with increased efficiency, should bring solar cell array units within reach of most users, with costs equating those of more conventional systems, in the 2 to 4 cents per kilowatt hour range.

ARTISTS CONCEPT OF A SPACE STATION FOR THE PRODUCTION OF ELECTRICITY -

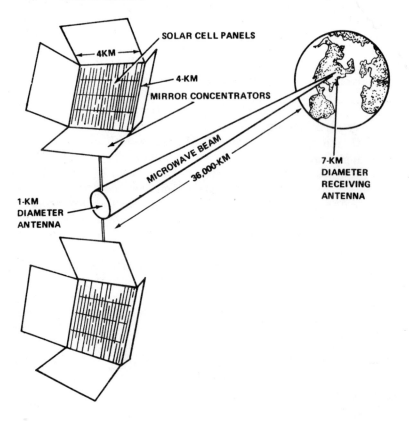

Of the three systems mentioned previously, the first, installing solar cell array systems on buildings is probably the nearest to a reality at this time.

The big plus factor is that the generator is installed at the place of use, which eliminates the need for an outside delivery system, and matches collected energy with actual need. Since it's virtually impossible to store more than an average day's needs, some type of auxiliary power would be required. As the concept is now planned the solar cell array would be mounted, or in some cases integrated into, a building, in combination with flat plate thermal collectors to complete a package including water heating, space heating, air conditioning, as well as electrical power. The illustration on page 34 shows the cross section of such a collector. The schematic on page 35 illustrates the components of a complete energy package. This is, of course, for a commercial application.

Before you rush out to buy components for a solar system, there are some additional things to consider. First, how well insulated is the structure. If the building in question is not well insulated, a solar energy package may be a waste of money. This is applicable to retrofit (adding a solar package to an existing building not built primarily with a solar package in mind) more so than an O.E.M. (original equipment plan). If you are planning the home specifically for solar energy, make sure it is well insulated.

Secondly, make certain the solar energy plant is large enough to accomodate the structure. Since installing a solar package is quite an investment, don't skimp on components. You will save money in the long run.

Thirdly, check zoning laws, environmental impact studies for your area, and your legal status regarding potential obstacles (such as a tree, tall building or other similiar structure) that might reduce the effectiveness of your system.

Finally, shop carefully. Check components for potential effectiveness. Be sure the unit is capable of meeting your requirements.

CROSS SECTION THERMAL–PHOTOVOLTIAC COLLECTOR

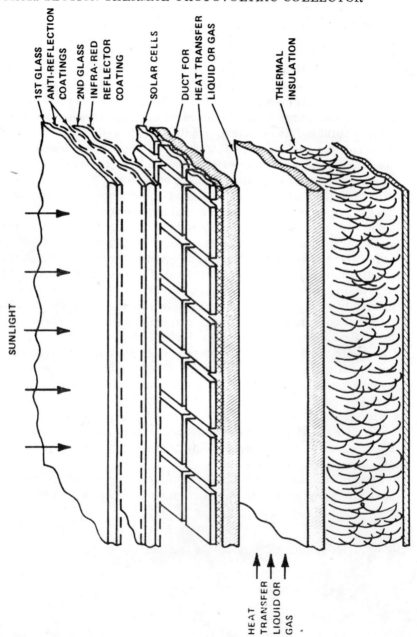

One version of a total package involving the use of Solar Energy

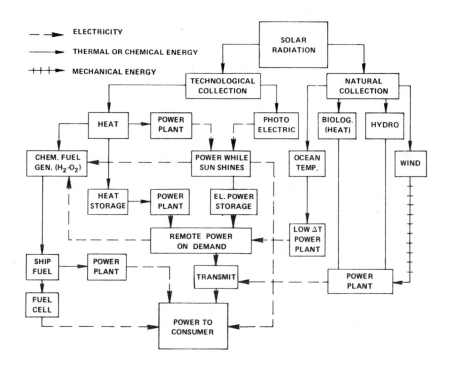

Schematic of a solar energy system for a residence.

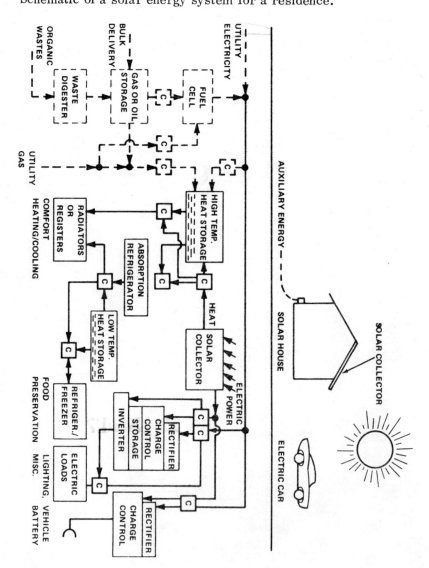

CHAPTER THREE

HEATING YOUR SWIMMING POOL WITH SOLAR ENERGY

As solar energy technology advances, one application very much in vogue is for heating the swimming pool. Most pool heating, to date, has been with conventional means using natural gas, fuel oil or electricity, but recently, another method, solar energy, has been rapidly gaining favor. There are several reasons for this, including among other things, a concerted effort in many parts of the United States to curtail the use of natural gas or oil for heating swimming pools. As this is written, legislation is under study in both Arizona and California which will prohibit future gas hookups to swimming pools. Although such laws are vigorously opposed by the pool industry, the handwriting is on the wall, and the ban will eventually come to pass. This will most certainly enhance the use of solar energy for pool heating applications.

The big obstacle to the use of solar energy is the initial cost. In most cases, it is considerably more than for conventional gas, oil, or electrical hookups. However, the amortization time on a solar unit is about five years-this includes reduced utility bills-and after that the operating cost is almost nil. And unlike natural gas, oil, electrical heaters, maintenance costs are very low. The reason is simple, there is no fuel, no additional pumps(in most cases) and no additional electrical controls. This is especially true where the unit can be connected to the existing water lines, and pump/filter systems. The energy source, the sun, is free.

As with conventional pool heaters, the solar heater can extend the swimming season from three to four months longer(depending upon where you live), and in the Southwest or zone 1 on the map on page

20, it means, for the most part, year-round swimming. In the case of flat plate collector systems it can also be utilized to reduce pool temperatures during the heat of the summer. Turning the system on after dark will aid in the dissipation of excess pool heat, and without appreciable chlorine loss as is the case with most swimming pool aerators.

Some questions that arise when discussing solar energy for pool heating include: "How efficient is the system?", or "How much will the temperature rise?", or "Will the system change the aesthetics of our home?" These questions are, of course, relative, but in most cases, the answers will be on the plus side. For example, regarding efficiency, a pretty safe rule of thumb is that in good, direct sunlight, the water temperature should rise from 10 to 16 degrees. The important thing to remember is that even on cloudy or hazy days the system should still function at from 65% to 80% of efficiency. And if the pool is covered between periods of use, look for another 5 to 10 degree rise.

Refer again to the map on page 20. It is obvious that certain areas are more conducive to solar energy than others, but the fact remains that at least 90% of the United States can benefit from a solar energy system with a large degree of efficiency, and that all of the country can benefit somewhat from a solar energy unit.

As for aesthetics, placing collectors on south-facing roofs should not change the appearance too much. It might be a good idea to place some kind of screening around the area to reduce visibility from the street or your neighbor's home. Other ideas, such as placing the collectors on the ground, or incorporating them as patio covers or as window awnings, are also possible.

Check with the local zoning department before installation to avoid conflict with local zoning laws.

Relative costs have been mentioned frequently, but it must again be emphasized that over a long period of time, the solar unit should save money over the use of conventional fuel or electricity. And this has a two-fold advantage since it helps save natural resources and fossil fuels, and also allows the pool owner to use the pool during periods of power cutbacks or fuel shortages. On balance, there is no reason why the largest percentage of new pools, or existing pools retrofitted with a solar unit shoudn't be heated by the sun. It is still possible that the solar heater may soon be the only option as more stringent rules governing fossil fuels are passed.

The preceding paragraphs relate primarily to the flat plate type system, and the use of solar discs, blankets or mattresses, which are placed directly on the pool's surface can provide even greater savings. Removal and replacement of the units is the biggest drawback, especially with the solar mattress which is quite cumbersome.

The floating air mattress concept has been used quite successfully in Australia, where tests revealed reduced evaporation and radiation losses increased pool temperatures by as much as 10 degrees fahrenheit.

Solar discs are easier to handle, but are somewhat less effective because they don't cover the entire surface.

Solar Discs on a Swimming Pool

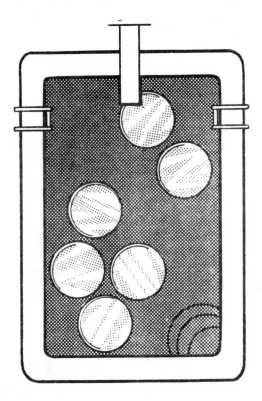

The most popular method of heating a swimming pool with solar energy is the flat plate collector method. This system utilizes the equipment already in the average pool setup to pump the pool water through the pool filter, into the collector modules, where the water is heated and piped back into the pool. The key component in this system is the collector (collectors). Refer back to chapter two for an in-depth look at various collectors. Collectors work best when mounted in a south-facing configuration, and roof-mounted units can be best placed for maximum efficiency.

The typical collector for a pool heater is constructed in modular form which features either heat collecting piping (generally copper) or water channels (such as corrugated sheet steel) which are fed by the feeder pipe from the pool pump and filter. The surface where the water flows is painted black to retain heat, and when the water is heated, it flows back into the pool. The most satisfactory flow would be to circulate the entire contents of the pool at least once daily. Pool size and capacity determines total collector area and how many modules are required.

A variation of this concept is used by Fafco Heater Company, of Palo Alto, California. Their concept features a sandwich-like collector (it resembles the cross section of corrugated cardboard), in which water flows evenly through the inner surface, then out the feeder pipe into the pool. The entire collector is formed of plastic which makes a much neater appearance.

CROSS SECTION OF A FAFCO COLLECTOR SYSTEM

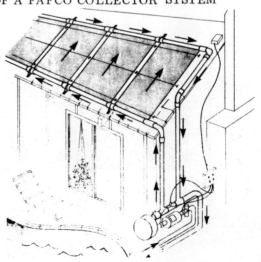

The modular collector used in heating a pool is somewhat simpler than that required for space heating or cooling. It doesn't require the same level of efficiency because it isn't as necessary to attain the higher temperature required in space heating operations. One component almost absolute in space heating, but not required for pool heating, is a layer of insulation between the collector modules and the roof of the house. We're referring here to retrofit units. The use of the glass plate cover is also optional, however, the use of a cover plate will not only increase efficiency, but will reduce water evaporation and keep foreign matter out of the system. This is especially important if pool water flows openly through outside channels. As mentioned above, the plate is not necessary for the system to function since pool water is generally colder than daytime temperature so there is little heat loss by convection.

The following system is manufactured by Sunergy Pool Heaters, and is typical of modular collector type that is both quite efficient yet relatively inexpensive. It functions much the same as a home radiator heating system, in reverse. Water is circulated through the pool pump/filter system onto the collector modules - generally located on the roof, but can be in any suitable location - and back into the pool after heating.

A collector module system is so versatile it allows the pool owner to use as many modules as required to adequately heat a pool. One module, consisting of two component panels 4 feet by 6 feet, for a total module size of 4 feet by 12 feet, should be used for each 100 square feet of pool surface area (see diagrams on following page).

The formula can be easily translated: If the pool size is 20' x 30', the surface square footage is 600', and the required number of modules would be 6 each of the 4' x 12' for a total collector area of 288' or roughly 1/2 of the pool surface square footage. This is based on the assumption that a 10 - 15 degree temperature rise is adequate. If your pool is located in the northern part of the country, where more heat is required, more modules must be added.

Module costs may vary somewhat according to materials used, and also if modules are purchased as a complete package or fabricated from components. There will be other collector types covered later in this chapter.

There are instances where the only system required may be the black polyethelyne pipe system. In this plan, black pipe is looped back and forth on the roof or other structure. This method is much less expensive, but is also less efficient, unless large amounts of piping are used.

THE SUNERGY SYSTEM

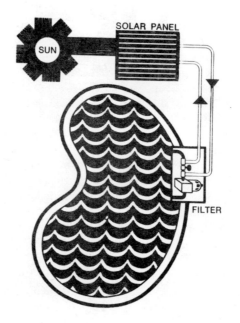

Page 43

Another method utilizes corrugated sheet steel as a base for the collector. In this system the water flows through the channels into feeder trough and back into the pool. To lay out this system, place galvanized sheets into frames and establish a flow pattern between all the plates. Place a feeder pipe across the top and drill holes adjacent to each flow channel in the plate. Attach a feeder trough at the lower end to catch the heated water and divert the flow back into the pool. The returned water should also be filtered before it returns to the pool. The water flow rate should be set between 200-300 gallons per 100 square feet of collector panel. Pump during daylight hours only. Although the system won't be quite as effective it can be operated during slightly overcast days. However, it should be shut down during stormy, rainy or snowy, and heavily overcast days.

The biggest selling feature of this system is cost. It is very inexpensive, and if greater efficiency is required, glass plating can be installed over the collector.

THE SOLAR BLANKET

The solar blanket is another inexpensive method used in heating a swimming pool. The concept is quite simple and involves a pool-sized blanket spread over the entire area. Pool water is the energy absorber. It traps the incoming heat and absorbs it as the blanket prevents escape. Additionally, a pool blanket cuts down on evaporation. This type of heater is capable of raising the heat level from 10 to 15 degrees, depending upon where the pool is located, time of the year, and amount of direct sunlight.

A typical Cases Flat Plate Collector-Face on view-

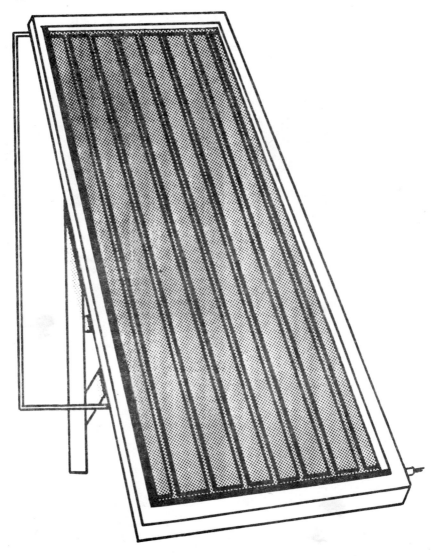

A typical pool heating system utilizing solar energy.

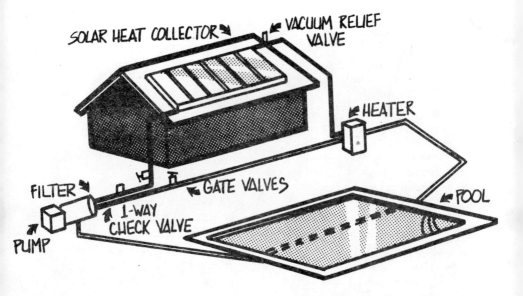

CHAPTER FOUR

SOLAR WATER HEATERS

Hot water for family use with the sun as an energy source is one of the most advanced applications of solar energy, both in the U.S. and abroad in such countries as Israel and Australia. The use of solar water heaters had it's beginning in Florida during the early 1920's and gained in popularity to a peak just prior to World War II.

The early units vary only slightly from today's models. A typical system contained a flat plate collector and a galvanized steel tank, connected by pipe and operated on the thermosiphon principle, for water circulation in the collector to tank and return loop. The tank was placed above the collector which permitted the less dense hot water in the collector to rise to the top of the tank and also prevented reverse cooling circulation during the night.

The big difference in those early models and most of those featured today, is improvement in collector efficiency, and the addition of controls which enables the system to be piped into a conventional fuel-fired water heater, thus reducing fuel consumption in the existing water heater. Multiple collector arrays can increase the rate of efficiency to a point where a conventional unit is rarely activated.

An operable solar water heater can be easily constructed by anyone with average mechanical ability and the cost is only determined by which components are used. We will feature one relatively inexpensive model using corrugated steel, and others that use copper tubing, and cost more to fabricate. As in other applications, the collector is the most costly component. The more efficient the collector, the more costly the unit. Items such as double pane glass, which definitely raise efficiency level, thicker insulation between the collector and roof, and insulation of connecting pipe and the tank, are a factor in system performance.

Diagram of a typical Roof-Mounted Solar Heater in Florida

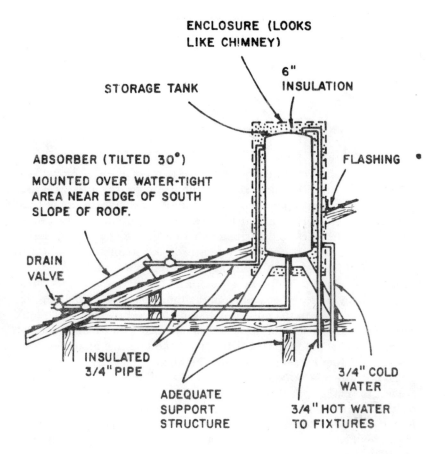

Typical Ground-Mounted System-Florida

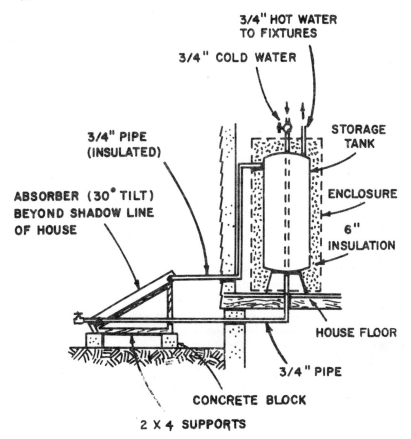

Sample absorber tubing configurations:

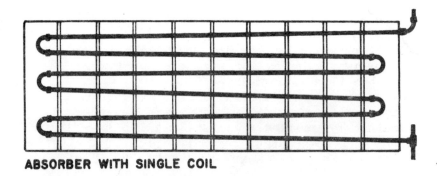

ABSORBER WITH SINGLE COIL

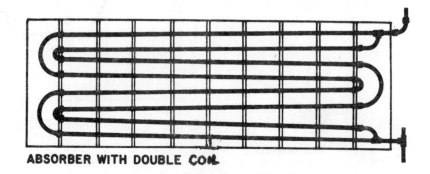

ABSORBER WITH DOUBLE COIL

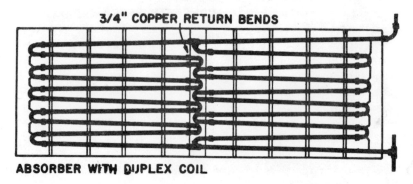

ABSORBER WITH DUPLEX COIL

Another difference between early models and those of today, well not really a difference, since the same concept is used today, but when a solar heater was installed in Florida, at least until around 1941, it was the primary system, with a booster for backup. It is not widely known, but one expert from the Florida solar water heating industry estimates there were once more than 60,000 units in Southern Florida, mostly in the Miami area.

They were installed in new homes and retroffited on older homes. And, as is the case today, retrofitting was more expensive and less satisfactory, mostly from an aesthetics point of view, because it's very difficult to avoid the "add-on look".

There were three basic places for positioning collectors, and the same is true today. The roof was most popular, with the ground-mounted unit, or as an awning over windows, also utilized.

The figure on page 48 illustrates a roof mounted unit. This was a retrofit unit - most systems constructed when the house was built were integrated as part of roof construction -with the storage tank placed above the topmost portion of the collector, which makes for unassisted (by pump or mechanical means) circulation of the water by gravity. The tank is well insulated and placed in a specially-constructed fixture(generally a metal structure designed to appear like a fireplace chimney).

Both water lines from the collector were inclined slightly to speed water circulation, and the line connecting the collector top and the tank was insulated to prevent reverse circulation during the night. The drain valve was strategically located so it could drain the tank and collector at the same time. A periodic draining was required to clean the system and get rid of corrosive buildup.

The figure on page 49 illustrates a system positioned next to the house on the ground, with tank located inside the wall. Although it was less expensive and easier to maintain, the collector was subject to damage and possibly less effective if trees or shrubs were planted nearby. Note that the same configuration applied, with the tank located above the uppermost portion of the collector. Note also that the collector is erected beyond the maximum shadow line of the house.

The use of collectors for window awnings isn't a bad idea and could be quite functional by killing two birds with one stone. It really does not change house appearance too much, and it shades south-facing windows, thus preventing heat buildup in the home. A disadvantage could be that the size could restrict collector efficiency, unless two or more collectors are used.

The most widely used types of collectors used during those halcyon days in Florida were copper tubing soldered to a copper base. The three most commonly used types are illustrated on page 50. Note that the copper tubing is not layed out parallel but on a slight angle. This helps speed water circulation through the system. The double coil and duplex types (middle & lower, page 50) are utilized to speed recovery time when all of the hot water is drawn off during peak usage periods.

The duplex coil was manufactured by the Solar Water Heater Company of Miami, Florida, and they claimed the design circulated 20% more water and produced water up to 30% hotter than absorpers with single coils.

The base material was originally very light copper and as designs improved, increased to from 6 to 10 ounce stock. Tubing was then, as it is now, mostly 3/4" soft copper. The space between each layer was about 6" in most cases. Some manufacturers placed the copper tubing closer together, but the only thing it really did was increase the cost, since tests proved that the 6" spacing produced 93% of the efficiency gained in closer spacing. Another interesting statistic is that if 3 ounce copper base material was used instead of 6 ounce the tube spacing had to be reduced to 4" to gain equal efficiency. Continual soldering of the tubing to it's base provided much greater efficiency than if spot soldered, since spot soldering provided less heat flow. This loss could, however, be compensated for by less space between the tubing.

The actual collector component is encased in a weathertight frame (see the cross section of a typical unit on page 54). The first units made by Solar Water Heater Company were constructed of 24 gauge galvanized steel but rust was a problem, so they changed to a unit comprised of 1/8" galvanized angle iron and asbestos, which wasn't satisfactory, so they finally settled on 18 or 20 gauge galvanized, sheet steel. Many units were made of this material, but the unit finally used was fabricated of 1/8" galvanized angle iron with an aluminum sheet.

Refer to the drawing and note the presence of insulation. There were different types used including mineral wool, regranulated cork or cork board, insulating wall board, saw dust, or vermiculate. The insulation serves a two-fold purpose, to prevent heat losses through the back of the collector and to protect the roof from any collector heat.

The base sheet of copper and the tubing were painted with a dull, flat paint(black is most widely used today) which provided adequate absorption of the sun's energy over the life of the unit.

The number of glass cover plates used varied from one to two, depending on location, with two glass plates used in areas subjected to freezing, although the second plate also prevents some heat loss and makes a more efficient unit. The glass type was window glass, single strength, supported on a wooden strip(see page 54). It was weatherstripped at the joints and held in place by a 16 gauge, galvanized iron, holding strip. The edges of glass were caulked, all around, to form a tight seal.

The storage tanks were standard, commercial-type, galvanized steel, with supports of either steel or wood(see drawings on pages 48 & 49). Insulation was used-sizes 5 to 6 inches - and the unit was then encased in a chimney-like affair on the roof. If the system was ground-mounted, the tank was installed just inside the wall nearest the collector component.

Cold water entered the tank through a line near the bottom, or in some cases through the tank top down to about 6" from the base of the tank (see illustration on page 49). The pipe carrying hot water to the tank exited near the top of the absorper(collector) and was piped to the top of the tank, which allowed the hot water to be drawn from the top.

The heater referred to here, and true of most systems, operated on the thermosiphon theory for circulation of water in the collector-tank circuit. Hot water is less dense and rises to the top of the collector as it heats and continues on to the tank as cold water returns from the tank bottom to the collector bottom to start the reheating process again. The ideal separation between the collector top and tank bottom was found to be two feet. The main reason for this is to stop reverse circulation when the sun isn't shining. The addition of insulation on the pipe from collector top to tank bottom also aided in that regard.

Larger installations often used a pump to force water circulation, especially when the tank was located lower than the collectors. In installations of that nature, a check valve was used to prevent the reverse circulation. Many of the systems manufactured today contain similar controls, including photocells.

Cross section of a typical collector illustrating components

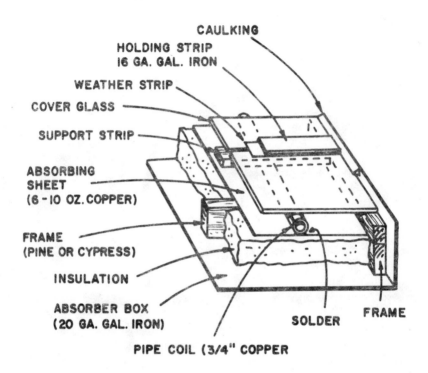

The circulating pump is activated by electric impulses from those slow-acting photocells, which are designed so that a passing cloud will not shut the pump off. Another concept currently in use starts the pump by sensing the temperature difference between the tank and the collector.

Correlation of tank size to collector area is an important factor in attaining maximum efficiency. If the collector is too small to heat the tank (we covered this earlier in swimming pool chapter), the water will never be hot enough. If the opposite is true, the water will be too hot and the collector efficiency will be diminished. The studies conducted in the Miami area indicate a well engineered unit will produce 1.5 to 1.7 gallons of water, at 130 degrees fahrenheit, per square foot of collector. The table below illustrates some of guidelines used in the Florida survey.

Residences

No. of Bedrooms	Occupants	Size of Tank (Gallons)	No. and Sizes of Collectors (Double Coil Type)
1 or 2	3	66	One: 9 ft. - 10 in. x 4 ft.
2 or 3	4	82	One: 12 ft. - 3 in. x 4 ft.
3 or 4 (2 baths)	4	100	One: 14 ft. - 8 in. x 4 ft.
Small duplex	6	120	One: 15 ft. - 11 in. x 4 ft.

Apartments

No. of Units	Size of Tank (Gallons)	No. and Sizes of Collectors (Double Coil Type)
3	200	Two: 14 ft. - 8 in. x 4 ft.
6	300	Three: 14 ft. - 8 in. x 4 ft.
10	500	Five: 14 ft. - 8 in. x 4 ft.

There are many factors which determine the effectiveness of the flat plate collector system. Heaters using the thermosiphon concept can produce temperatures as high as 165 degrees fahrenheit when using the double coil system (refer to page 53) according to tests in the Miami, Florida, area. Variables include latitude, which also determines tilt and angle degree of collector, number of glass plates and, of course, size of collector relative to tank capacity for the temperature of delivered water. The table below illustrates degree of effectiveness in the Miami study.

TABLE 2. ABSORBER ARRANGEMENTS AND CAPACITIES FOR HEATING WATER TO 130° F. IN FLORIDA*

Location	Angle of Tilt from Horizontal Facing South	Number of Cover Glass Layers	Water Heated from Air Temperature to 130° F. (gallons per sq. ft. per day)	
			Minimum Season	Annual Avg.
North Florida (Jacksonville, Gainesville, Pensacola, etc.)	45°	1	1.0	1.5
		2	1.1	1.7
	30°	1	0.8	1.3
		2	1.0	1.7
South Florida (Miami, etc.)	45°	1	1.4	1.6
		2	1.5	1.7
	30°	1	1.4	1.7
		2	1.6	1.8

Maintenance then, as now, on the solar water systems was relatively nil. It was recommended that the entire system be drained and cleaned at least twice yearly. The drain valve was located so both the tank and collector could be drained simultaneously. This periodic draining helped keep rust down to a minimum. It was also a good idea to clean the glass plate and check all seals and frame for leaks, and correct as needed.

After some time, many solar water heater owners added boosters as back up energy source. Most systems contain such an innovation today, or in another version of the same concept pipe the solar unit into the existing hot water heater, which reduces the use of the conventional fuel system as long as the solar furnished water maintains a certain temperature. Most units added in that Florida survey, were gas fired boosters added to the top of the tank (to avoid heating the entire tank). The primary reason for adding boosters at that time was the sudden proliferation of new appliances which required more hot water and overtaxed the existing solar unit. The boosters of today are thermostatically controlled, generally with a manual on-off switch so the booster could be by-passed, to cut electrical consumption. This would be a good idea during long periods of hot, sunlit days and at vacation time.

COLLECTOR USED AS AN AWNING

Another example of an efficient, yet inexpensive solar water heater is one built by the Brace Institute, McGill University, Quebec, Canada. The original system was built and tested in Barbados, West Indies. Several test models over an extended period of time have proven very efficient.

As stated previously, solar water heater components are the most readily available for general use, not only here in the United States but Australia, Israel, Japan and Russia and other countries as well. Solar water heater components can be expensive, they can be inexpensive, or they can be moderately expensive. The following unit is quite inexpensive. It is constructed of very basic materials, and the life expectancy of the system will be very nearly that of a much more sophisticated design.

This unit was specifically designed to be constructed of simple, low-cost materials available most everywhere (note that the pilot project was located in the West Indies), including remote areas of under-developed countries. The heater is designed to provide from 30 to 40 gallons of hot water per day ranging in temperature from 130 to 140 degrees fahrenheit (depending upon local conditions).

At the time of the original experiment the cost was approximately 45.00 American, local purchase, for all components including: the collector, hot water storage tank, cold water storage tank, and the connecting piping. Labor costs were not included, and this would be applicable on a do-it-yourself project here as well. If the rate of inflation and local costs are now considered, this unit could still be built most anywhere in the United States for under $100. The life expectancy is at least five years with minimal maintenance, and is likely much longer. The water storage units (oil drums for the hot water tanks) would probably go first, so replacement costs would be a minor item. Maintenance consists of draining and cleaning the entire system twice annually (rust buildup is the big problem) and an occasional cleaning of the unit.

Refer to the following pages and note required components consist of four main parts:
1. The collector/absorber which absorbs the sun's rays, and heats the water.
2. The collector casing which houses the absorber, and insulation materials.
3. The hot water tank-suitably insulated-to hold heated water.
4. The cold water feed tank (also made from an oil drum).

GENERAL SCHEMATIC OF SOLAR WATER HEATER

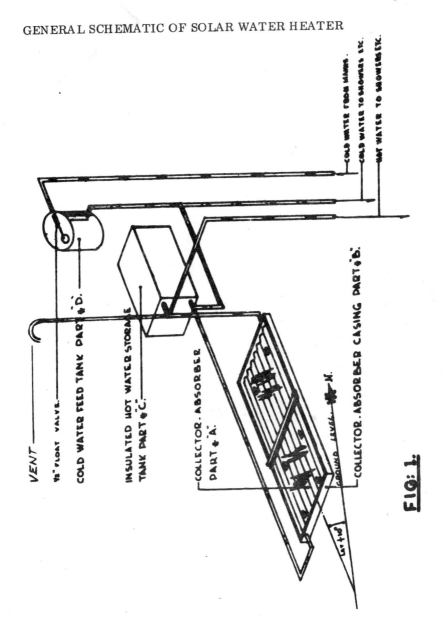

FIG: 1.

ELEVATION OF SOLAR WATER HEATER SYSTEM

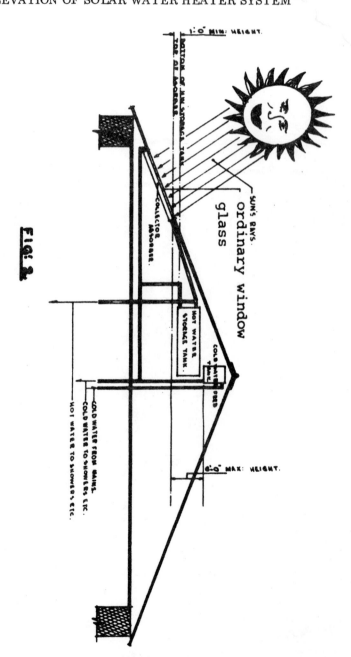

Fig. 2.

BUILDING THE COLLECTOR

The absorber is constructed of standard galvanized steel, 22 ga. 26" wide by 96" long, with corrugations 3" apart and 3/4" deep. A second piece of flat galvanized steel, slightly larger all around, is also required. Refer to material list and schematic-pages 62 & 63.

1. Cut the corrugated sheet to 26" x 88 5/8" with a pair of the metal shears-they can be purchased from sheet metal shop.
2. Cut the flat sheet to 26 1/2" x 90 5/8", also 22 gauge.
3. Place a sheet of stiff cardboard against end of corrugated steel, and trace shape of corrugations with a black pencil. Cut along the traced line for later use as a pattern. Place the cardboard pattern on each end of the flat sheet steel-part(#2), figure A-and mark the corrugations.
3a. Cut along the marked lines with metal shears. The sheet ends should now resemble sketch Z, page 63.
4. Cut two holes, 0.84" in diameter(one in each end per sketch) to allow future placement of a length of 1/2" pipe.
5. Attach a 9" length of pipe (1/2" diameter) to each end of the corrugated sheet(#1) as shown in sketch X. Screw the 3/16" 1/4" screw into the galvanized pipe and solder the pipe to the corrugated sheet steel.
6. Bend both ends of the sheet steel (2) at right angles per the sketch W. The bent ends should be 1" long(1/4" longer than depth of corrugations to allow for some overlap when soldering. Place the corrugated sheet(#1) on top of the sheet(2) and slip the 1/2" pipes (#4) into the corresponding holes, but it will be necessary to cut the sheet(2) at the end(per sketch Y) to allow the pipe to enter the hole.
7. Bend the edges of the sheet(#2) over the corrugated sheet(1) as shown in sketch V, and solder as shown. To best bend the edges of the flat sheet over 1/4" from edge as needed, clamp the sheet between two pieces of angle iron along the edge to be bent and strike with hammer to get a right angle.
8. Drill 1/4" holes for rivets in the inverted corrugations, as shown in sketch U. Place 1/4" rivets (#3) in the holes with the heads resting on the flat sheet, as shown, and peen the rivet heads. Solder over the rivet heads to complete construction of the collector absorber.
    a. Test for leaks by placing the unit in a sloping position, by standing up against a building, and fill with water not un-

Materials List

Part No.	No. Off	Material	Size
A		**- The Absorber**	
1	1	corrugated galv. steel sheet	22 gauge, 8ft. x 26in.
2	1	"special" flat galv. steel sheet	22 gauge, 8ft. x 36in.
3	28	galv. steel rivets	¼in. dia., approx. 5/16in. long
4	2	galv. steel water pipe	½in. I.D., 9in. long
5	2	m.s. machine screw	3/16in. dia., ¼in. long
6	2	sticks of solder	
B		**- The Absorber Casing**	
7	1	"special" flat galv. steel sheet	24 gauge, 8ft. x 3ft.
8	8	galv. rivets for ends of casing	¼in. dia., approx. 5/16in. long
9	-	coco-nut fibre or equivalant insulation	20 lbs.
10	6	22 gauge galv. steel sheet	1in. x 1in. x ¾in., supporting "L" brackets
11	6	felt strips or suitable insulation	1in. x 1in. x ⅛in. thick
12	6	galv. rivets for part (10)	¼in. dia.
13	4	22 gauge galv. steel sheet	1in. x ¾in. x ½in., hold-down "L" clamps
14	4	galv. steel self threading screws for part (13	⅛in. dia. x ½in. long
15	1	22 gauge glav. steel sheet	27⅛in. x 2½in., to make glass support rib
16	4	glav. rivets for part (15)	¼in. dia., approx. 5/16in. long
17	2	sponge rubber strip, e.g. "Dor-Tite"	¼in. x ⅛in. x 17¾in. long
18	1	sponge rubber strip, e.g. "Dor-Tite"	¼in. x ⅛in. x 22ft. long
19	2	window glass	27¾in. x 44¾in. x ⅛in. thick
20	1	silicone type sealant (or equivalent)	12 oz. cartridge
21	1	black plastic electrical insulating tape	one roll, 1in. wide (or nearest)
22	16	22 gauge galv. steel sheet	1in. x ¾in. x ¾in., hold-down "L" clamps
23	12	sponge rubber strip, e.g. "Dor-Tite"	¼in. x ⅛in. x ¾in. long
24	16	galv. steel self-threading screws	⅛in. dia. x ½in. long
C		**- Hot Water Storage Drum**	
25	1	used steel oil drum	standard size 45 gallons
26	2	"special" flat galv. steel sheet	24 gauge, 8ft. x 4ft.
27	-	coco-nut fibre or equivalent insulation	30 lbs.
28	2	deal wood boards	1in. x 12in. x 9ft. long
29	1	heat resistant paint	1 pint tin
30	-	diesel oil	1 pint
31	-	gasoline (petrol)	1 pint
D		**- Cold Water Feed Tank & Piping**	
32	1	used steel oil drum	standard size 45 gal., ½ only required
33	1	plumbing float control valve	½ in.
34	-	½in. galv. steel pipe and fittings	to suit particular installation
35	1	Rust-Oleum or similar paint	1 pint tin

PARTS BREAKDOWN FOR THE ABSORBER/COLLECTOR

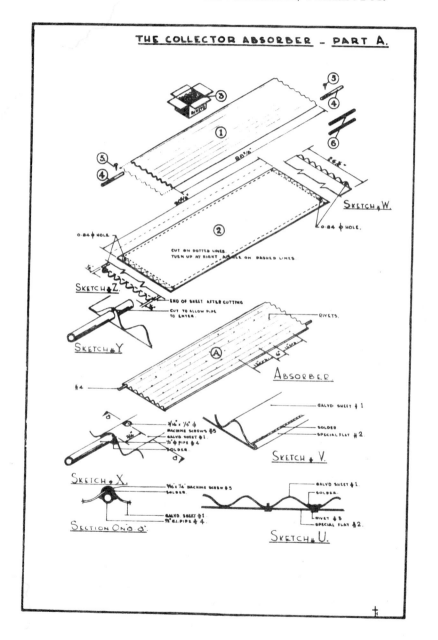

der pressure from a water system, and allow it to stand for some time. Observe area of leaks and mark with chalk. If the unit has some slow leaks and repair procedures doesn't stop them, leave them and they'll no doubt seal themselves in time. After the leaks are repaired, allow the unit to stand in the sun, filled with water.
   b. Paint corrugated side with two coats of flat black paint.

BUILDING THE ABSORBER CASE - Illustrated on page 65
1. The base of the absorber case is also made from flat galvanized steel, 24 gauge. Start with a sheet 3' x 8', and cut to 33 3/8" x 96", and round off the corners per sketch S, page 65, and bend the sheet at right angles along the dotted lines shown in the sketch. After the sheet is bent, sketch T, rivet the four corners with 1/4" rivets(total of eight rivets).
   a. Drill 2 1/2" drain holes in one end of the case, as shown in sketch T. When final assembly is complete, the holes are at the lower end of unit for condensed moisture outlet. holes for later installation to case with rivets.
2. Cut part(#10) which consists of 6 each L-brackets, and drill 1/4" holes for the rivets in the brackets.
3. Bond 1/8" thick felt strips(#11) cut 1" x 1" to L-clamps.
4. Check sketch T and drill 6 each 1/4" holes in the case. Install L clamps(#10) onto case with flat head rivets(#12) with head facing to the outside.
5. Place insulation into bottom of case (sketch T & V), per the material list, use a 2" layer of whatever insulation you want.
6. Place the absorber into the case over insulation but resting on the L-supports, and with black corrugated surface facing upward.
7. Make 4 each L-clamps(#13) from 22 gauge steel-size 1/2" x 3/4" x 1". Use self tapping screws (#14) and screw the L-clamps in place (per sketch T) to hold absorber to the case.
8. Make a T-rib per sketch T, and form it from a sheet of 22 gauge steel(it's possible to buy such a rib from certain ceiling tile wholesalers). Make it the same width as the case, and rivet the rib to case with 2 galvanized rivets each end.
9. Place the 1/4" Dor-tite or whatever rubber stripping you are using on the rib(#15) per sketch U.
10. Place 1/4" wide stripping all around the 1/4" edge of case.

THE ABSORBER CASE-PARTS BREAKDOWN

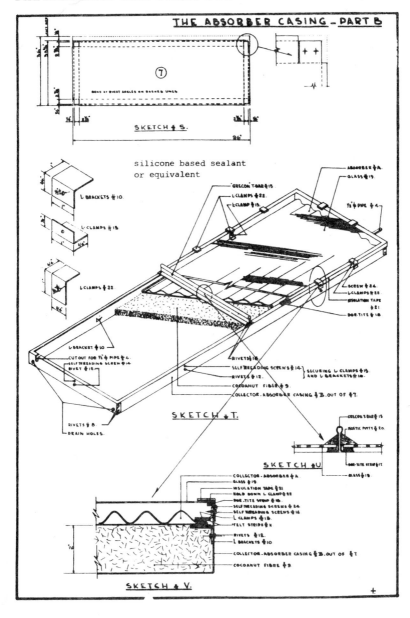

11. Place the two sections of window plate glass, size 27 3/4" x 44 3/4" x 1/8" thick(19) into place on the case, and make sure they rest evenly on the stripping, all around the case.
12. Apply silicone sealant between the glass and center and the edge supporting ribs. Allow 1/8" on each side for expansion of glass. Ordinary putty can be used but silicone sealant lasts longer. Seal the glass to the case with black electrical tape wide enough to cover the seams.
13. Make 16 each L clamps (#22) 3/4" x 3/4" x 1" from the 22 gauge steel; drill a 1/8" hole in each. Place 3/4" sponge rubber strips(#23) on the glass over the tape, and attach the L clamps in places indicated on sketch T & V. Press clamps over the rubber strips and drill a hole into the case, 1/8", and insert the self-threading screws (#24) into the clamps, until all are in place. This completes the collector unit.

THE HOT WATER STORAGE DRUM

1. Use a 45 gallon drum - or a similiar size as available - and rinse the interior with a half pint of diesel oil (to dissolve any oil left in the drum, then rinse again with a half pint of gasoline to dissolve the diesel oil, which should leave the drum grease free, and dry enough to paint.
2. Pour a pint of heat resistant paint(#29 materials list) into the drum(#25) and shake thoroughly until interior of the drum is coated with paint. Place drum aside and allow it to dry.
3. To build a wooden frame for the oil drum, refer to drawing on page 67.
4. Cut frame according to dimensions on the drawing.
5. Insert the oil drum into the framework and cover the frame with sheets of 24 gauge sheet steel (#26 materials list), but leave the top open so you can fill the empty spaces with the pieces of insulation.
6. Place reducers in the two oil drum holes, and attach one length of galvanized pipe 6" x 1/2" in each opening. After the pipes are in place, fill remaining spaces with insulation and attach top of frame in place.

COLD WATER FEED TANK

1. Cut a standard oil drum in half(#32 materials list) and install a 1/2" float valve(#33) as shown on page 59.

TANK FRAME CONSTRUCTION

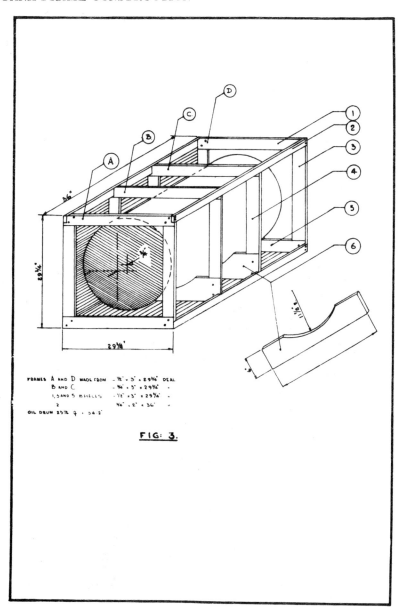

FIG: 3.

2. Paint the inside of the drum with two coats of Rust-Oleum paint and allow it to dry.
3. Obtain lengths of piping as needed to suit your application, and connect the system per illustrations on pages 59 & 60, and the following instructions.

ASSEMBLY OF COMPONENTS

1. Use pipe sealant on all connections before tightening the final time. Make all connections in elevations noted on pages 59 & 60. Note especially the 1' difference between the top of the collector and bottom of the hot water tank.
2. The collector must face the south (or north, if south of the equator at an angle equal to the latitude of the area, plus 10 degrees.
3. The layout will be more functional if piping is kept to short distances. If for some reason the hot water drum cannot be placed adjacent to the collector, use 1" pipe instead of 1/2" If this is not done, the hydraulic resistance of the piping may be so high that natural convectional circulation of the water between absorber and tank will be badly restricted. And the system's efficiency will be severely hampered.
4. Finally, allow the unit to function for at least one day before drawing hot water. This will give the system enough time to heat up. Be sure there is enough water in the system at all times. Too little water can create abnormal heat in the unit and cause considerable damage. If it becomes necessary to empty the system (i.e. to clean it, or winterize it), cover the collector with a tarp.

COLLECTOR WITH HORIZONTAL TUBING

The following collector is similar in design to those used in the Florida Solar Water Industry except the tubing is placed up and down on the absorber plate instead of lateral.

The tubing can be either copper or aluminum, but a corresponding metal must be used for the base since dissimilar metals used together will cause corrosion. Again, the tubing should be soldered all along contact point with the base for maximum effeciency of heat transference.

ILLUSTRATION OF COLLECTOR WITH VERTICAL RISERS

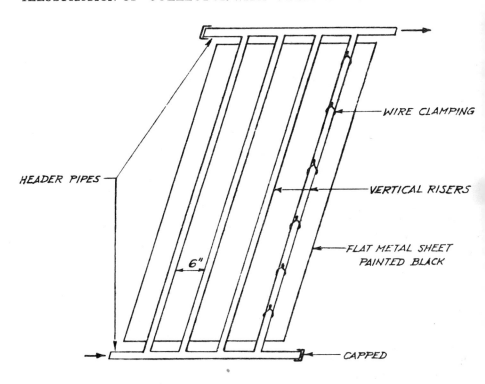

CHAPTER FIVE

SOLAR COOKING & A SOLAR STILL

One of the most widely used methods of solar cooking is making a solar tea(sun tea) and requires no additional appliance. Place the tea in a gallon jar and set the jar in south-facing exposure. The tea will be brewed in a very short time.
There are a variety of solar cookers ranging from a simple foil, generally aluminum, shaped in a parabolic circle, to cookers which contain solar cells to enhance the heating factor. It is possible to boil water in 20 minutes, even with a simple cooker.
Solar cookers enjoy wide popularity around the world, mostly in Australia and Asia.
The following plans are for a solar cooker which utilizes a separate collector box for heating the water.

HOW TO MAKE A SOLAR COOKER
This cooker is a combination of two basic components which are permanently joined. First, the solar collector which heats water and the second is the steam cooker in which the pot or pan containing food is placed.
The collector is placed in the ground at a fixed angle of 45 degrees, on a pipe base. The collector is set so it faces the sun in the morning, and can be turned to face the sun as it travels across the sky. This versatility allows a unit to be positioned to take full advantage of the sun, but is generally not required unless semi-cloudy, or cloudy conditions prevail.
The collector must always contain water, and water should be add-

ed each evening to replace that which has boiled away. Under normal conditions, steam should be produced within an hour after the sun first strikes the collector, and will continue being produced throughout a typical sunny day. This makes it possible to cook as many as three meals a day, with a minimum of both a mid-day and evening meal. In fact, food remaining in the cooker will stay hot for several hours after sunset.

Since solar cooking is a slow process, many crock pot recipes or recipes that require long cooking will be the best bet. They would include stews, vegetable dishes, and even hot cereals.

Other uses of a solar cooker are varied. A unit can be assembled in your back yard barbeque area where the cooked meal can either supplement the barbeque cooking for suffice as the main course. A solar cooker is so versatile it can be used on camping trips.

The most practical application is for less fortunate people around the world, and in these days of awareness regarding world-wide hunger, the solar cooker can-and in many areas already does- go a long ways toward solving the hunger problem. And finally, it can be a valuable asset for the back-to-the-land advocates.

PLANS & FABRICATION (by Brace Institute, McGill University)

Refer to the materials list on page 72 and obtain all the necessary components. Paint all wood pieces that make up the framework. Use a good white undercoat paint. Do not paint the hardwood panels.

1. Assemble collector case, per drawing page 73, and install support cross brace (#A11), and the two long support strips (A4), down the sides, to support the copper base. There is no support at the ends. Affix the upper end brace(A10), inside of the box.
2. Wet the rear panel of tempered hardwood for several hours then nail that panel(#A3) in place.
3. Drill clearance hole for supporting pipe at 45 degree angle, into 2" x 4" x 7" block of wood, then fasten in place with the screws for that purpose. Screw them down from inside.
4. Drill holes in each end board to accomodate the 3/4" water tube(B1). Drill in the middle of end board about 1 1/2" from bottom of the board.
5. Paint the outer box, sides and ends, a dark color such as black, or a dark green.
6. Set the collector case(box) aside and assemble the absorber.

MATERIALS LIST FOR COOKER

Materials List for Cooker

Part No.	No.	Name and Size
Wood		
A 1	2	Box sides ½" x 4" x 62"
A 2	2	Box ends ½" x 4" x 20"
A 3	1	⅛" tempered hardboard rear panel 62" x 21"
A 4	2	Mounting strips for copper plate 1" x 2" x 61"
A 5	2	Spacers ¾" x ¾" x 61"
A 6	2	Spacers ¾" x ¾" x 18¼"
A 7	2	Frame for inner glass sides ¾" x ¾" x 60⅞"
A 8	3	Frame for inner glass, cross pieces ¾" x ¾" x 18¼"
A 9	1	Glass support strip, see drawing ¼" x 20" x 1½"
A 10	1	End brace (for attaching support arms for steam bath) 1" x 2" x 16"
A 11	1	Support cross brace 1" x 3" x 20"
A 12	1	Pivot block 2" x 4" x 7" (with 45° hole for pivot pipe)
A 13	1	Bottom skid 1" x 2" x 30"
A 14	2	Support arms for steam bath 1" x 2" x 24"
A 15	2	45° wedges to support steam bath 1" x 10" on long face
A 16	2	Steam-bath platform cross-members 1" x 2" x 9"
A 17	1	⅛" hardboard 9" x 10"
Miscellaneous		
B 1	1	¾" pipe 65" long, threaded both ends
B 2	1	¾" cap, screwed ¾" BSP (British Standard Pipe thread)
B 3	1	¾" 135° pipe spring
B 4	1	¾" pipe backnut
B 5	1	1" i.d. x 4" long rubber tube
B 6	1	0.025" copper sheet 20" x 59"
B 7	1	Insulation 20" x 61" x 22"
B 8	1	Aluminum foil 18" x 61"
B 9	1	Pivot pipe ½" x 12"
B 10	1	Glass ⅛" thick x 30" x 72"
B 11	1	Glass ⅛" thick x 30" x 72"
B 12	1	Steam bath fabricated from 26 gauge, 22" x 60" galvanized sheet, see drawing
B 13	1	Lid, fabricated with B 12
B 14	1	Hinge, 9" long, galvanized
B 15	1	Saucepan with lid, 8" dia. 5" deep
		White undercoat paint
		White gloss paint
		Flat black paint
		Any other dark paint
		1/16" galv. wire, 10 ft. long

CROSS SECTION OF THE SOLAR COOKER

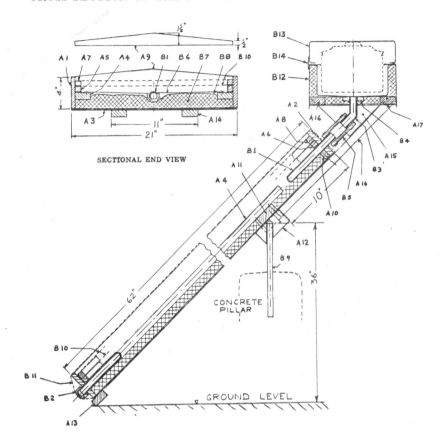

COPPER PLATE AND PIPE ASSEMBLY

Refer to drawing on page 73, and attach pipe(B1) firmly along the centerline of the copper plate (B6). Do this by partially wrapping the plate three-fourths of the way around the pipe, then drawing the open ends of the plate together over the pipe, by using galvanized wire at 3" intervals along the length of pipe. To make a better heat transference system and for a better bond, it would be a good idea to solder along the seam. However, if the copper plating is tightly wrapped around the pipe, this may not be necessary.

The pipe should extend 2" past the copper plate at the lower end, and 4" past the plat edge at the upper end(per drawing page 73).

Place insulation material(B7) in the 2" space in the box base, and cover with layer of aluminum foil(B8). Any loose fibrous material may be used for insulation, but it should be packed quite tight.

Insert the pipe-plate assembly and nail plate to the side mounting strips(A4). Screw the bottom end cap(B2)into place(to be removed only for cleaning out corrosion and rust from the system).

Paint the plate with a light coat of flat black paint, but be sure the plate is clean, and after the first coat is dry, add a second coat to aid absorption rate.

Add the glass plates to the collector box after assembling framework(A7, A8) which should fit loosely into the box. Before placing glass in, insert spacers A5 & A6-size 3/4" x 3/4" and install with inner plate glass-use size as listed in the materials list.

Add the center spacer(A9) and suitable frame for holding a second pane of glass. Seal glass edges with a silicone-type sealant.

FRAME FOR STEAM COMPONENT

Assemble the support frame (A14, A15, A16)for the steam bath, see figure on page 73, and paint the frame gloss white. Screw to underside of the collector box.

Cut the hardwood underpiece(A17)to fit, and drill a hole in center to accomodate the pipe.

Construct the steam bath(B12) and solder the bottom pipe backnut (B4) in place.

When constructing the steam bath, be sure the horizontal parts of the unit slope slightly to the front so all water and condensate will drain back into the pipe from the collector.

Rivet the hinge (14) of lid (B13) to outer case. Fill the 1" space around the steam bath with insulation.

STEAM BATH COMPONENTS–SOLAR COOKER

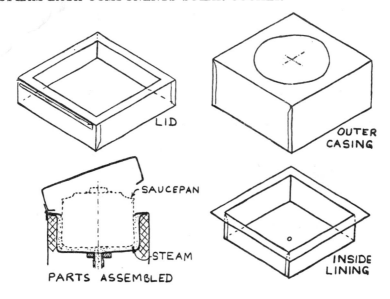

DETAILS OF STEAM BATH

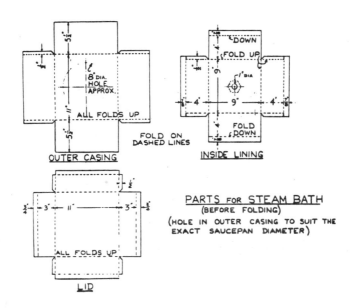

PARTS FOR STEAM BATH
(BEFORE FOLDING)
(HOLE IN OUTER CASING TO SUIT THE EXACT SAUCEPAN DIAMETER)

PARTS FOR STEAM BATH

Refer to the parts list and screw the 135 degree pipe spring (B3), into the backnut under the steam bath assembly. Be sure it doesn't protrude up into the bath. Place insulation into space under the bath and place the unit in position. Use a short length of radiator hose to join the two lengths of pipe, and wrap entire area of exposed pipe with insulation.

Paint the outer surface of the steam bath a dark green or black.

When the unit is assembled and ready for use, the bottom of the cooking pot should stand about one half inch above the inside lining, which will allow boiling water to pass freely into the bath from the connecting pipe. The case should be constructed so side handles of the pot rest on the top of the outer case of the bath. If this isn't possible, place a lightweight wooden or metal pot stand(similar to those used in pressure cookers) in bottom of the steam bath. A pan of aluminum would be the best type to use.

The pipe for holding the collector case should be installed so the top of the pipe is 36" above the ground(see figure on page 73). The pipe should be 1/2" diameter, and can be set in concrete.

It may be necessary to use a platform since the top of the unit is about 5 feet above ground.

AN EXPERIMENTAL SOLAR ENERGY UNIT

The following solar unit consists of components to form a 2' x 4', sheet metal base, with 1/2" copper tubing soldered to the base-in the pattern illustrated page 78 and a wooden frame covered by glass or clear plastic. Note the similarity to units covered in the water heater chapter.

As in previous text, the storage tank should be well insulated, as should the pipe which connects the collector to the tank. Note also that the tank is elevated above the uppermost portion of the collector as in previous diagrams. One feature of this unit is the lack of a pump to force the water from the collector into the tank. This not only cuts costs, but reduces reliance on conventional power sources.

This unit operates on the theory of thermodynamics, since water like air, expands and becomes lighter, as the temperature rises. The cold water enters the lower pipe, becomes warmer and lighter and rises through the lower tubing to the upper tubing and into the outlet pipe leading into the tank, where cool water returns to the unit for reheating.

The collector should be south-facing at a latitude reading for the area, plus 10 percent. A 30 degree tilt should be adequate for an effecient operation.

On the test unit, in the midwest, the starting water temperature at 7:00 AM was 60 degrees. By 5:00 PM that afternoon, the tank temperature had risen to 128 degrees. Overnight, the water had cooled to 98 degrees, but on the second day when the ambient outside temperature rose to 70 degrees, the water temperature in the tank had increased to 138 degrees, or almost double the ambient temperature. On subsequent days, the temperature of the water ranged form a low of 108 degrees(on a cloudy day) to a high of 148 degrees, and all of this without the assistance of any outside energy source.

This sample of solar energy on a small scale tends to support the theory that widespread use is not only feasible, but viable.

If you refer back to the chapter on solar water heating you will see that findings on this experimental model parallel the test results in the Florida survey.

The source for this information is MOTHER EARTH NEWS.

Refer to the illustration on the following page.

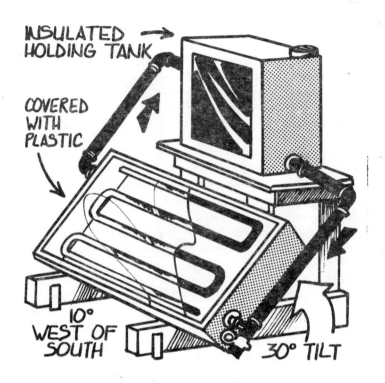

THE SOLAR STILL

Distilling water with solar energy is certainly nothing new. It is probably the oldest use of the sun's rays. Portable distillers were used to purify sea water for human consumption for some time, and during World War II, fliers used a similar kit to good advantage. Residents of ocean fronts have utilized the sun to remove salt from sea water, and conversely, salt has been mined by a reverse usage of the same process.
Many years ago, the University of Arizona built a huge plant for desalinization of sea water on the Gulf of California, near Puerto Penasco (known to Anglos as Rocky Point) Mexico. Although much larger in scope, the basic principle is the same as in the smaller systems. As a result of the pilot plant in Mexico, the University of Arizona later built a similar plant in the shiekdom of Abu Dhabi in the Middle East.
A very basic solar still can be constructed of three pieces of glass cut in a triangle and glued together. The glass is then placed over a pan of saline water over an insulated pad. When placed in direct sunlight, the sun's rays will create water condensation, which will be carried to the top of the unit by air circulation. The water is then funneled off into a catch-tray. As the process continues, all water will be condensed, leaving only a salt residue in the pan. As stated above, the principle is similar to most salt mining where sea or other heavily saline water is pumped into ponds and allowed to dry. When the water evaporates, only the salt remains. It can then be mined as needed.
The drawing on the following page is our artists conception of a more sophisticated system, and is quite capable of distilling several gallons of water per day.

ARTISTS CONCEPTION OF A SOLAR STILL

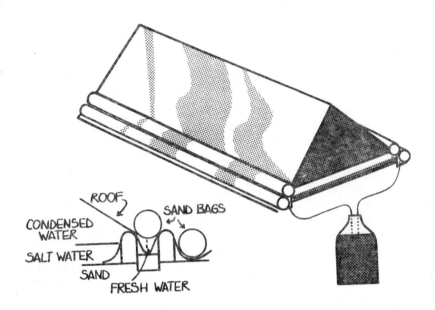

CHAPTER SIX

SPACE HEATING AND COOLING

The previous chapters have covered some of the projects open to the average do-it-yourselfer. This chapter will deal with space heating and cooling in the context of experimental homes and public buildings in which systems have been installed for observation, as well as those designed for functional application.

When I first started researching this book there were perhaps two or three dozen such projects around the country. As I complete the book almost three years later, the applications are becoming quite commonplace. In fact, many builders around the country are now incorporating solar systems into new housing units. One of those builders is Ernest Carreon of Tucson, Arizona and his Consolar solar energy package will be covered later in this chapter. There is every indication that using the sun for space heating and cooling is on the increase and we hope it continues. If more contractors take the plunge and do a good selling job, the message might filter back to Washington.

This last chapter will also go into some detail on that pilot project in Timonium, Maryland. In my estimation the project proves beyond a doubt that solar energy applications in public buildings should be considered. The installation in that school was expensive, but it was a retrofit which required modification to the existing structure and engineering costs were much higher than would be the case if a solar energy system was installed when the building was erected. Additionally, it was installed while the school was in session, which added to the cost. The Timonium project was the first to be installed in a school and preliminary studies indicate it will save money over an extended period of time. The collector array contains 5100 sq.

feet of collector area and the huge storage tank of 15,000 gallons provides 3 gallons of water per collector feet. It's a huge installation, but much data should be derived from the numerous instruments that are programmed to record all facets of the operation.

The pilot unit was installed in the middle section of a three section building (see illustration on the following page). By using a single section the engineers are able to determine relative effectiveness compared to the sections not utilizing a solar energy system. It has simplified solar system studies, and preliminary figures do indicate the unit is quite effective (NSF bulletin #ER7934 available from the United States Government Printing Office covers the entire project in detail). In fact, the study documents the savings of about 1200 gallons of fuel oil in a period from March 14 to May 15, 1974, in which the solar heating system provided 91% of the center wing's heating requirements for that period. The following chart indicates efficiency of the system on a given day. It shows how the efficiency rises and falls with insolation (for a given water temperature).

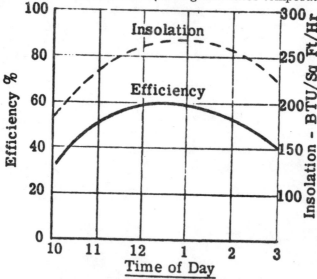

A.M. Inlet water 140 degrees Fahrenheit
P.M. Inlet water 147 degrees Fahrenheit
Outlet @ 156 degrees Fahrenheit average
Temperature Ambient 70 degrees average
Date: April 29, 1974- Conditions, no wind and clear

SOLAR ENERGY EXPERIMENT–Timonium Elementary School

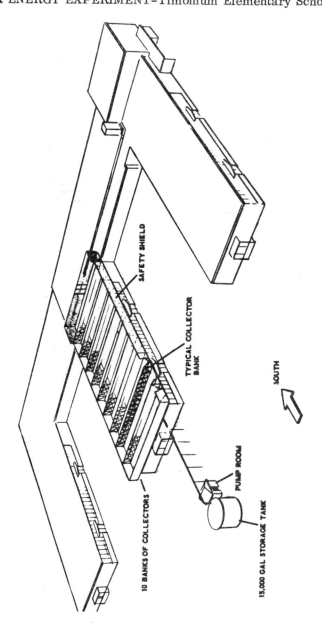

Illustration of basic solar system-Timonium Elementary School

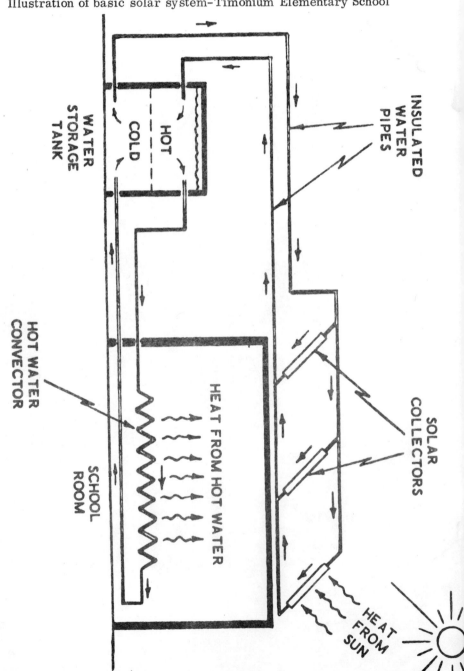

The collector array was made up of 180 units 4' x 7' x 2-3/8" with each unit weighing approximately 90 pounds. See illustration below.

The collector was constructed of two glass covers made from low-iron content, double strength glass. The sbsorber plate was made of aluminum and two thicknesses of aluminum honeycomb filled the area between glass and absorber plate. The honeycombing was then notched before extended in place, to allow for passage of the water.

Both glass plates, the honeycomb, and absorber plate were bonded together with two-part epoxy and a 1 1/2" thick sheet of polyurethane foam, sheet insulation was bonded to the back of the absorber. After the foam edging was applied, a rubber strip was added to seal the unit. See illustration below.

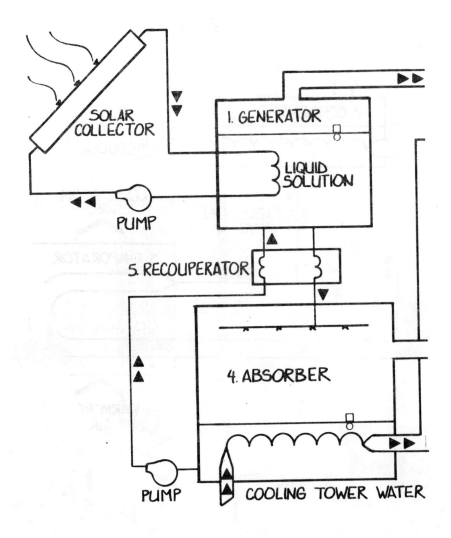

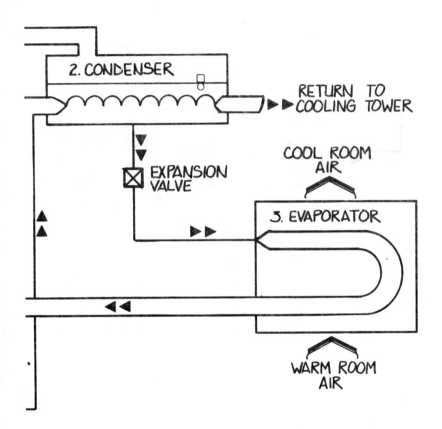

The drawing on page 86-87 illustrates one system that can be used for space cooling. It is an absorption program whereby solar heat is utilized to cool inside room air.

Refer to diagram which shows how this absorption system works. The components are; generator(1), condenser(2), evaporator(3), and absorber(4). The fluid used in the system is a solution of refrigerant and absorbent which have an affinity for each other. Water is the refrigerant and lithium bromide is the absorbent.

When the solution in the generator is heated by solar energy from the collector, the water (refrigerant) is vaporized form the liquid solution. The water boils off(vaporizes at less than 212 degrees F. because the pressure in the generator is less than normal atmospheric pressure). Heat is removed from the vapor in the condenser and the vapor changes to a cooled liquid, which is then revaporized through an expansion valve, and passes into the cooling coils of the evaporator. Vaporization of the refrigerant takes place due to lower pressure beyond the expansion valve relative to pressure in the condenser.

Warm room air blown over the cooling coil of the evaporator is cooled and the cool air is then circulated into the rooms. By vaporizing the refrigerant, the temperature is lowered from 120 degrees to 70 degrees fahrenheit, a difference of 50 degrees(the temperature of the vapor refrigerant at the evaporator is about 70 degrees lower than the temperature of the liquid refrigerant) which makes it more effective in cooling the room air.

The next step finds the vaporized refrigerant recombined with the absorbent in the absorber. Heat is generated in the recombination and heat is removed by cooled water from the cooling tower. The temperature in the absorber must be low enough to assure a high-chemical affinity between refrigerant and solution. Finally, the solution from the absorber is returned to the generator and the cycle is started again, and repeated continuously.

The recouperator(3) is used to cool the high temperature solution in the generator as it passes into the low temperature absorber, and to heat the low temperature absorber solution as it moves to the high temperature generator. This serves to minimize heat loss associated with fluid transfer between the absorber and generator.

The thermodynamic properties of the working fluids in the absorption system are such that absorber temperature and concentration of the solution in the absorber determine pressure in the absorber.

Absorber pressure, which is the same as evaporator pressure, also determines the temperature at the evaporator and the amount of effective cooling possible. The temperature of the condenser determines it's pressure, which is the same as the generator pressure and generator pressure and the concentration of the solution determine the temperature required to vaporize the refrigerant (water). Therefore, the temperature at the absorber and condenser, and the concentration of the solution determine the minimum temperature in the generator(i.e. the temperature of cooling water from the cooling tower - approximately 75 degrees for the experimental home as tested by Colorado State University - and the concentration of the refrigerant-absorbent solution, determine the minimum temperature in the generator which must be supplied by solar collectors before the system will function (approximately 180 degrees in test house). Another consideration is a temperature reading at the recouperator for a given concentration of solution beyond which the system will encounter problems due to crystallization(approximately 120 degrees in the test house).

The test house was constructed with NSF funding by Colorado State University, Fort Collins, Colorado. The following text covers other facets of the operation in diagram and explanation.

This experimental house is known as Solar I and was constructed in the Rocky Mountain area.

The purpose was to design an effective, yet economical, system of heating, cooling, and heating domestic water in a typical residence using primarily the sun's rays. The heating and cooling system was designed to handle three fourths of the heating and cooling requirements while maintaining a conventional level of comfort using a fully automatic control system. The remaining one fourth would be supplied by a backup system utilizing conventional methods. The house is a modern three bedroom home with 3000 square feet of heated space on two levels.

HOUSE SPECIFICATIONS
    Total floor space: 3000 square feet
    Floor area, main level: 1500 square feet
    Floor area, Lower level: 1500 square feet
    Roof pitch: 45 degree from horizontal
    Design heating load: 17,600 Btu/degree day
    Design cooling load: 26,000 Btu/degree day
    Insulation: Ceiling-5" fiberglass batt/Wall: 3 1/2" fiberglass batt.

Page 90

The house was located at 41 degree N, latitude, and 105 degree W longitude, at an altitude of 5200 feet.

The heating/cooling system is divided into five basic units; The solar collector array (1), solar storage (2), house heating, cooling and domestic hot water load requirements(3), conventional furnace auxiliary system(4), and automatic controls(5).

Solar collectors make up a total array area of 768 square feet and are mounted on a south-facing roof at a 45 degree angle from horizontal, which allows for optimum collection on a year-round basis.

As we've noted previously, the best collector arrangement would be one in which the collector always directly faced the sun, but this is impossible, unless the collectors are programmed to track the sun across the sky. The 45 degree angle is based on the best position on a year round basis.

As with collectors covered previously, the solar energy passes through the glass plates and is absorbed by the metal plate. See diagrams below. This system is somewhat different than those covered previously since collector fluid (water) contains a 25% solution of ethylene glycol (anti-freeze).

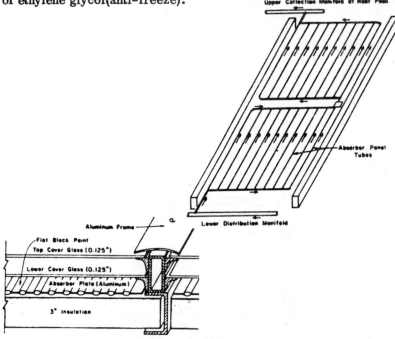

The collector fluid is pumped from the basement into a lower distribution manifold, upward through the collectors, and into the upper manifold (see illustration on page 91). Heated fluid leaves the upper manifold and returns to the basement where heat is transferred to the storage unit (see figure below).

Since ethelyne glycol is very expensive, the liquid is not piped into the storage tank, but is piped through a heat exchanger (see figure below). The system uses only about five gallons instead of the large amount (about $1000 worth) that would be required.

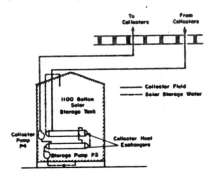

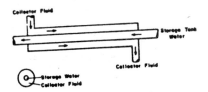

The absorber plate in this collector is aluminum plating with the tubing placed in the pattern per drawing on page 90. The cross-section drawing on page 90 also illustrates component makeup of unit. The plate is painted with flat black paint to attain the highest degree of absorptivity. The unit also contains 2 glass plates made of double strength window glass, 1/8" thick.

The research team on this project decided to use an 1100 gallon storage tank. A smaller tank would inhibit domestic water usage, and a larger tank would require a larger collector system, and the cost would be considerably higher.

The heat distribution system used in this test house is by forced air. Their choice was determined by heating and cooling requirements and the availability of commercial cooling units. The system operates when hot water from the storage tank is piped to the heating coil or directly to the cooling unit(figure on page 93). Air that is blown across the heating coil picks up heat to be carried to the house.

For cooling, hot water from the storage tank is piped to the air conditioner where it provides the necessary energy to operate the cooling unit. The unit is a 3-ton lithium bromide absorption unit converted for use with hot water instead of natural gas(see drawing on page 86-87 and text on page 88).

In addition to heating and cooling, the system also provides most domestic hot water(see figure page 93). Water from the cold water main enters the pre-heat tank and is circulated through a heat exchanger. Hot water from the solar storage tank transfers heat to the domestic hot water. As needed, water from the pre-heat tank enters a conventional gas hot water heater which can raise the water temperature if required, and maintain the water at the desired temperature. An auxiliary system is included as a backup unit if the solar system doesn't supply the necessary energy.

If the temperature in the storage tank drops below a predetermined point (100 degrees F for heating, 170 degrees F for cooling), the auxiliary boiler automatically takes over and provides heat to the heating coils or the air conditioning system. Under normal operating conditions, the solar unit will provide 75 to 80% of heating and cooling load, and the gas-fired backup system, 20-25%.

The control system is somewhat more complicated in this home than in a conventionally heated home, since it must also control the collector and storage pumps and two automatic valves in addition

Page 93

to those normally required. This control system is fully automatic with full system control capability. All that is required, is a proper thermostat setting.

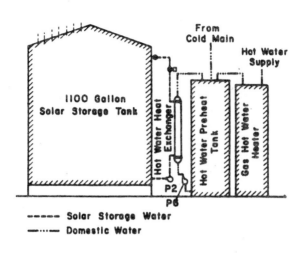

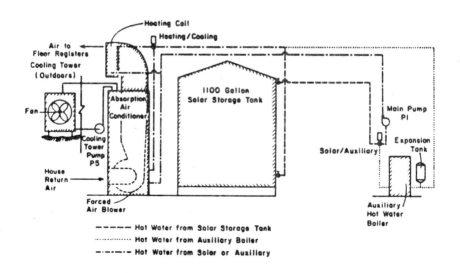

INDEX

Absorber 25, 26, 63, 65
Alternative Energy sources 6-18
Alternative type collector 69
ARKLA 8
Arizona Public Service 15
Arizona, University of 10, 14, 15
Atomic Energy Commission 8
Australia, 6, 7
Brace Institute, McGill University 58, 71
Cadmium sulfide solar cells 31
California, 15, 18, 21
Cells, solar, 31, 30, 34
Colorado State University 86-93
Consolar system 94
Decade 80 solar house, 94
Edmund Scientific 31
ERDA 9, 10, 15, 21
Faquier School 10, 11
Flat Plate Collector 24, 25, 27, 45, 50, 54, 57, 69, 89-91
Flat Plate Collector-thermo-photovoltiac 34
Florida 7, 12, 21, 54-56
Ford, President Gerald R. 21
France, 6, 7
Heating-Solar Energy 81, 85-94
Heating & Cooling 85-94
Heating Water 47, 48, 49, 59-65, 87-93
Honeywell 8, 19
HUD(Housing & Urban Development) 9
Insulation 16-18
Israel 6, 7
Japan 6, 7
Meinel, Dr's Aden & Marjorie 14
Mother Earth News 78
NASA 30, 9
NSF 9
New Mexico 11
Olin 8
Ozone layer 5

PPG(Pittsburgh Plate Glass) 8
Parabolic collectors 25
Photovoltiac cells 30, 34
Ray, Dixy Lee 8
REA(Rural Electrification 6
Revere Copper 8
Silicon Cells 8, 30-31
Solar batteries 30-31
Solar Blanket 44
Solar cooking 70-
Solar Collectors 24, 25, 27, 34, 35, 45, 50, 54, 57, 89-93
Solar Discs 39
Solar energy meaning 5
Solar Heating & Cooling Demonstration Act 1974 9
Solar energy conversion to electricity 35-36
Solar Energy system 23, 86-87
Solar One House 86-93
Solar Stills 70, 79
Solar Swimming Pool Heaters 37, 42-46
Solar Space Station 32
Tax breaks 17
Texas 15
Thomason, Harry 12
Thermal-photovoltiac cells 34
Timonium Elementary school 11, 28-29, 81-85
Truman, President Harry S. 13
Tucson Gas & Electric 15
Tyco Labs 30
United States 6, 7, 20, 37
USSR (Russia) 6, 7
Water Heating 47-49, 59-63, 65, 89-93
Westinghouse 8
Wind 6
Zoning 15, 33